Modern Neurosurgery of Meningiomas and Pituitary Adenomas

Edited by

R. Fahlbusch, W. J. Bock, M. Brock,
M. Buchfelder, M. Klinger

Acta Neurochirurgica
Supplement 65

SpringerWienNewYork

Professor Dr. Rudolf Fahlbusch
Neurochirurgische Klinik, Universität Erlangen-Nürnberg, Federal Republic of Germany

Professor Dr. Wolfgang J. Bock
Neurochirurgische Universitätsklinik, Düsseldorf, Federal Republic of Germany

Professor Dr. Mario Brock
Neurochirurgische Universitätsklinik, Berlin, Federal Republic of Germany

PD Dr. Michael Buchfelder
Neurochirurgische Klinik, Universität Erlangen-Nürnberg, Federal Republic of Germany

Professor Dr. Margareta Klinger
Neurochirurgische Klinik, Universität Erlangen-Nürnberg, Federal Republic of Germany

© 1996 Springer-Verlag/Wien
Softcover reprint of the hardcover 1st edition 1996

Product Liability: The publisher can give no guarantee for information about drug dosage and application thereof contained in this book. In every individual case the respective user must check its accuracy by consulting other pharmaceutical literature.
The use of registered names, trademarks, etc. in this publication does not imply, even in the absence of specific statement, that such names are exempt from the relevant protective laws and regulations and therefore free for general use.

Typesetting: Thomson Press, New Delhi, India

Graphic design: Ecke Bonk

Printed on acid-free and chlorine free bleached paper

With 69 Figures

Die Deutsche Bibliothek – CIP Einheitsaufnahme

Modern neurosurgery of meningiomas and pituitary adenomas /
ed. by R. Fahlbusch ... – Wien; New York : Springer, 1996
 (Acta neurochirurgica ; 65)
 ISBN-13: 978-3-7091-9452-2   e-ISBN-13: 978-3-7091-9450-8
 DOI: 10.1007/978-3-7091-9450-8
NE: Fahlbusch, Rudolf [Hrsg.]

Cataloging-in-Publication Data applied for

ISSN 0065-1419
ISBN-13: 978-3-7091-9452-2

Preface

"Modern Neurosurgery of Meningiomas and Pituitary Adenomas" presents the state-of-the-art of neurosurgery for these two types of tumors. Following a classification of the pituitary adenomas according to pathology, molecular biological factors are presented and their effects evaluated as these aspects deepen our understanding of the growth and further expansion of these tumors. The diagnosis is made not only by a study of the hormonal status, but also by neuroradiology. A number of authors have devoted their efforts to the special problem groups such as the elderly patients and those with huge pituitary adenomas. Particular emphasis is of course placed on the surgical treatment, including transcranial and transphenoidal neurosurgery, but the use of medical treatment and irradiation must be discussed as well.

The treatment of meningiomas has also been influenced by the molecular biology of hormone and growth factors. Therefore it is accorded extensive space in this volume. The prognostic significance of nuclear DNA content is discussed. Recent research with new diagnostic methods such as somatostatin scintigraphy, PET studies end progesterone receptor in tumor fragment spheroids is presented here. The treatment of these tumors, however, depends largely on their localization. For the neurosurgeon, the surgical treatment of meningiomas involving the cavernous sinus, meningiomas of the ventral Foramen of Monroe, meningiomas of the cerebello-pontine angle and of the optic sheath is presented by prominent experienced leaders in this field.

Erlangen, March 1996

R. Fahlbusch
M. Klinger

Contents

Listed in Current Contents

Acta Neurochir (1996) [Suppl] 65: 1–3

Current Pathological Classification of Pituitary Adenomas

W. Saeger

Department of Pathology, Marienkrankenhaus, Hamburg, Federal Republic of Germany

Summary

A classification of pituitary adenomas basing on detailed structural and immunohistochemical studies is accepted world-wide and is mandatory for each pathologist. Monohormonal (densely or sparsely granulated GH cell adenomas, Prolactin cell adenomas, ACTH cell adenomas, FSH/LH cell adenomas, alpha-subunit-only adenomas), bihormonal (mixed GH/Prolactin cell adenomas, mammosomatotroph cell adenomas, acidophil stem cell adenomas), plurihormonal (GH/Prolactin/Glycoprotein-positive adenomas, other Glycoprotein-positive types) and hormone-negative adenomas (null cell adenomas, oncocytic adenomas) have to be differentiated.

Keywords: Tumor classification; pituitary adenoma; immunocytochemistry.

Principles of Classification

Pituitary adenomas can be classified in various ways. The classification depends on the morphological methods used by the pathologist. Staining characteristics such as acidophil, basophil or chromophobe are nowadays obsolete as main principle of classification because they do not identify special adenoma types. The growth pattern and the architecture are important but only additional factors for classification. DNA measurements can also give interesting information but do not identify special or distinct types as some adenomas are diploid and other adenomas of the same types are aneuploid [1]. Proliferation markers such as PCNA [5] or Ki-67 [3] demonstrate a rapid growth rate but do not identify a special adenoma type.

Very valuable information is available from immunocytochemistry for hormone content showing that about 60% of surgical specimens of pituitary adenomas contain one hormone, about 15% are bihormonal, 15% are plurihormonal and just less than 15% are hormone-negative. Immunocytochemistry is a good principle for classification but should not be used as the only one because it does not reflect the structure.

Most important are structural features for adenoma classification. Paraffin histology gives useful information for tissue pattern and architecture and in some cases also for cytological features. Thus calcifications are typical for Prolactin cell adenomas and spheroid bodies indicating fibrous bodies characterize sparsely granulated GH cell adenomas [4].

Better information is available from semi-thin sections of Epon-embedded material. They enable us to demonstrate clearly the densely arranged mitochondria in the oncocytic adenomas and the secretory granules especially of the GH-, Prolactin- or ACTH-secreting adenomas [7].

Superior for structural analysis is the electron microscope showing distinct structural differentiations especially if there are similarities with normal cells or not [4,7].

But there are also additional features characterizing special adenoma types: the partly pleomorphic secretory granules in the densely granulated GH cell adenomas, the fibrous bodies in the sparsely granulated GH cell adenomas, the two cell line-differentiation in the mixed GH/Prolactin-cell adenomas, the strongly developed rough endoplasmic reticulum and the misplaced exocytoses in the sparsely granulated Prolactin cell adenomas, the giant mitochondria in the acidophil stem cell adenomas, the bundles of type 1-filaments in ACTH cell adenomas, the elongated polar cells with smaller and larger granules in the gonadotroph adenomas, the poor development of cytoplasmic organelles in the null cell adenomas and the dense accumulation and increase of mitochondria in the oncocytic adenomas [2].

W. Saeger

Table 1. *Classification of Pituitary Adenomas (1991–1993)*

Adenoma type	Surgical series		Autopsy series	
	Frequency N	%	Frequency N	%
Densely granulated GH cell	26	6	2	1.5
Sparsely granulated GH cell	43	10	2	1.5
Mixed GH cell/prolactin cell	32	7		
Mammosomatotroph	10	2		
Acidophil stem cell	1	0.2		
GH/prolactin/glycoprotein	30	7		
Densely granulated prolactin cell	2	0.5	4	3
Sparsely granulated prolactin cell	59	13	45	32
Densely granulated ACTH cell	55	13	9	6
Sparsely granulated ACTH cell	27	6	10	7
Crooke's cell			1	1
TSH cell	1	0.2		
FSH/LH cell	33	7		
Glycoproteid hormone	12	3	3	2
a-subunit only			1	1
Plurihormonal unclassified	23	5	1	1
Null cell	44	10	39	28
Oncocytic	6	1	19	14
Not classified[a]	34	8	3	2
Total	438	100	139	100

[a] Due to insufficient material.

From light microscopy by paraffin sections, from immunohistology for all pituitary hormones and from embedding for semi-thin sections and ultrastructure the up to now best classification can be performed (Table 1). It depends on the structure and on the hormone content. So we differentiate two monohormonal GH cell adenomas: the sparsely and the densely granulated one, three bihornomal adenomas producing GH and Prolactin: the mixed GH/Prolactin cell adenoma which contains one GH secreting and one Prolactin secreting cell type, the mammosomatotroph with one cell type producing both hormones, and the acidophil stem cell adenoma producing more Prolactin than GH with one undifferentiated cell type. Frequently GH and Prolactin producing adenomas contain also Glycoproteid hormones, especially TSH and Gonadotropins. The tumor classified as adenoma containing other hormones is mostly a GH- and Glycoproteid hormone-positive adenoma without demonstrable Prolactin.

The monohormonal Prolactin cell adenoma is differentiated in a very frequent sparsely granulated type and a very rare densely granulated variant.

ACTH cell adenomas are also subclassified into a densely granulated and in a sparsely granulated type.

Monohormonal TSH cell adenomas are rare. FSH/LH adenomas are more frequent. Is an adenoma positive for FSH or LH in combination with TSH we designate it an adenoma, positive for glycoproteid hormones. An additional rare type is the a-Subunit-only adenoma.

The null cell adenoma makes up about 10% of the surgical material. The oncocytic adenoma can be interpreted as a variant of null cell adenomas.

Correlation to Clinical Functions

In acromegaly, about half of adenomas are monohormonal GH cell adenomas, 30% are GH and Prolactin secreting adenomas and 20% are plurihormonal types. Sellar GRH secreting gangliocytomas in combination with a GH secreting adenoma or just GH cell hyperplasias as a consequence of an extrasellar GRH producing, mostly pancreatic tumor are very rare [8].

In hyperprolactinemia, 80% of adenomas are monohormonal Prolactin cell adenomas, other types are rare. Mostly in cases with slight hyperfunction, tumors can be of inactive type or can be craniopharyngiomas or other tumors and the hyperprolactinemia is caused by paraadenomous Prolactin cell hyperplasias [6].

In Cushing's disease or Nelson's syndrome ACTH cell adenomas of the densely granulated or sparsely granulated type are found. Additional hormone contents are rare. Because many adenomas are very small and therefore the specimens contain only small groups of adenoma cells, the rate of those tumors which cannot be sufficiently classified is most of all high.

ACTH hyperfunction can be caused by ACTH cell hyperplasia, but those cases are very rare. If an adenoma could not be found, an extrapituitary origin of the hyperfunction should be looked for. Demonstration of Crooke's cells in the specimens is very important, since these cells are mandatory for hypercortisolism.

Isolated TSH hyperfunction is rare and caused by TSH cell adenomas, but TSH cannot be demonstrated in every case. FSH/LH hyperfunction is also rare. In those cases we find gonadotropic or glycoproteidhormone-positive adenomas. Just 46% of the clinically silent adenomas seem to be typical inactive adenomas and are represented by null cell adenomas or oncocytic adenomas. All others are adenomas of those types we also find as clinically active types. Very frequent are FSH/LH- and Glycoproteid-containing adenomas. We think that these adenomas are actively secreting types but do not increase the hormone plasma levels significantly. After stimulation with LHRH they mostly show increased levels.

Latent Adenomas

An interesting field is the adenoma in autopsy series which we should name latent adenomas because they generally do not show any clinical signs. Many but not all adenoma types can be found in those series (Table 1). Because most of them are still very small it seems that they represent a pre-clinical stage of actively hypersecreting adenomas.

References

1. Anniko M, Tribukait B, Wersäll J (1984) DNA ploidy and cell phase in human pituitary tumors. Cancer 53: 1708–1713
2. Horvath E, Kovacs K (1992) Ultrastructural diagnosis of human pituitary adenomas. Microsc Res Techn 20: 107–135
3. Knosp E, Kitz K, Perneczky A (1989) Proliferation activity in pituitary adenomas: Measurement by monoclonal antibody Ki-67. Neurosurgery 25: 927–930
4. Kovacs K, Horvath E (1986) Tumors of the pituitary gland. Atlas of tumor pathology. Washington: Armed Forces Institute of Pathology Sec Ser 21: 1–269
5. Mc Nicol AM, Sheperd M, Lane OP (1991) Cell proliferation in pituitary adenomas; correlation with hormonal immunoreactivity. Abstract 14. J Endocrinol Invest 14 [Suppl 1]: 55
6. Riedel M, Noldus J, Saeger W, Lüdecke DK (1986) Sellar lesions associated with isolated hyperprolactinemia. Morphological, immunocytochemical, hormonal and clinical results. Acta Endocr (Kbh) 113: 196–203
7. Saeger W (1981) Hypophyse. In: Doerr W, Seifert G, Uehlinger E (Hrsg) Spezielle pathologische Anatomie. Ein Lehr- und Nachschlagewerk. Band 14: Endokrine Organe, Teil 1. Springer, Berlin Heidelberg New York, pp 1–226
8. Saeger W, Puchner MJA, Lüdecke DK (1994) Combined sellar gangliocytoma and pituitary adenoma in acromegaly or Cushing's disease. Virchows Arch 425: 93–99

Correspondence: Wolfgang Saeger, M.D., Department of Pathology Marienkrankenhaus, Hamburg, Alfredstrasse 9, D-22087 Hamburg, Federal Republic of Germany.

Acta Neurochir (1996) [Suppl] 65: 4–6

Molecular Biological Research in Pituitary Adenomas from the Pathologists' View

K. Kovacs

Department of Pathology, St. Michael's Hospital, University of Toronto, Toronto, Ontario, Canada

Summary

Some recent findings related to pituitary adenoma pathology achieved by molecular biological methods are briefly reviewed. It is increasingly obvious that the application of the molecular pathology approach can provide a deeper insight into the causation, histogenesis, cellular derivation and differentiation as well as progression of pituitary adenomas and can help to understand better structure-function correlations.

Keywords: Immunocytochemistry; in situ hybridization; pathology; pituitary; pituitary tumors, ultrastructure.

Unprecedented progress was achieved in the last two decades in the better understanding of the morphology of the human pituitary and its diseases, primarily hypophysial adenomas. The introduction and extensive use of transmission electron microscopy, immunocytochemistry and immunoelectron microscopy provided a deeper insight into the cytology and pathomorphology of the pituitary. The separation of adenohypophysial cells and adenomas into acidophilic, basophilic and chromophobic types became outdated and was replaced by more meaningful cell and adenoma classification based on ultrastructural features, hormone content and structure-function correlation. One recently recognized important phenomenon was plurihormonality which rendered the one cell-one hormone theory, a dogma which dominated pituitary cytology for several decades, obsolete and paved the way to develop a more flexible and functionally oriented concept emphasizing the significance of inter-connection between various adenohypophysial cell types. In the last few years, the application of molecular biology and genetics became integral parts of hypophysial investigation providing a novel sophisticated level for increasing knowledge on cell develop-ment, differentiation, endocrine activity, regulation, oncogenesis and tumor progression. The rapidly expanding area of receptors helped to obtain a deeper insight into cellular function and markedly facilitated the design of new drugs which can be used effectively in the medical treatment of several patients harboring pituitary adenomas.

Pituitary adenomas are frequently occurring epithelial neoplasms originating in and composed of adenohypophysial cells (Kovacs and Horvath 1986). The light microscopic study of hematoxylin-eosin stained sections of formalin-fixed, paraffin embedded tissues remains the cornerstone of diagnostic pathology. Transmission electron microscopy is a valuable method in cell classification; it permits the investigation of the various steps of the secretory process, assesses endocrine activity, cell injury and drug effects. Immunocytochemistry is the most widely used technique in cell identification. It is sensitive, reliable and reproducible, it localizes hormones and many other constituents in the cell and can be applied also for autopsy material; tissues fixed in formalin and stored in paraffin for many years can be successfully immunostained. Immunoelectron microscopy is the application of immunocytochemistry at the ultrastructural level. It is time consuming, expensive but it is a valuable tool in antigen localization and cell recognition. Since two hormones can be labeled simultaneously in the same cell, immunoelectron microscopy is a rewarding procedure in establishing plurihormonality.

The demonstration of gene expression in the cells using in situ hybridization is a valuable tool; its application is expanding and it is used currently in many laboratories permitting the study of pituitary adenomas at the molecular level. In situ hybridization using oligonucleotide probes combined with S35 autoradiography makes possible the localization of various mRNAs in individual cells and connected with the streptavidin-biotin-peroxidase complex technique it documents gene expression and the gene product in the same cell. Oligonucleotide probes and a large number of antibodies are commercially available thus the presence or absence of various pituitary hormone-growth factor- and receptor-mRNAs in association with their gene product can be conclusively revealed in nontumorous and adenomatous adenohypophysial cells. Nonradioactive hybridization histochemistry offers considerable potential because in contrast to isotopic methods, it is less expensive and results can be obtained more rapidly.

Studies using hybridization histochemistry methodology and the literature focusing on molecular biology findings in human pituitary tumors were reviewed recently by Thapar et al. (1993).

One intriguing result is the demonstration of growth hormone releasing hormone (GRH) in pituitary adenoma cells (Joubert et al. 1989, Levy and Lightman 1992, and Wakabayashi et al. 1992). We have confirmed and extended these studies (Stefaneanu et al. 1994b). By in situ hybridization we have found GRH mRNA in several densely granulated somatotroph adenomas, sparsely granulated somatotroph adenomas, mixed somatotroph-lactotroph adenomas, mammosomatotroph adenomas, acidophil stem cell adenomas, functioning and silent corticotroph adenomas, thyrotroph adenomas, gonadotroph adenomas, null cell adenomas and oncocytomas. The presence of GRH mRNA was conclusively documented in somatotroph adenomas removed surgically from patients treated with octreotide, a long acting somatostatin analog. Lactotroph adenomas of patients treated with bromocriptine, a dopamine agonist, expressed the GRH gene. The majority of positive adenomas exhibited a diffuse, weak or moderate hybridization signal over the tumor cells; in some mixed somatotroph-lactotroph adenomas and acidophil stem cell adenomas GRH mRNA was unevenly distributed among adenoma cells. The presence of GRH mRNA could not be correlated with a specific hormone, morphologically classified adenoma type, or with tumor growth. It was intriguing that GRH mRNA was present not only in somatotroph adenomas but in all the other adenoma types. GRH is known to stimulate growth hormone secretion and somatotroph multiplication; its effect on the functional activity of other adenoma types remains to be elucidated.

Surface receptors and nuclear receptors play a major role in regulating cellular functions including hormone synthesis and release, cell multiplication and tumor growth. We have studied the presence of estrogen receptor mRNA using in situ hybridization in a large number of nontumorous and adenomatous adenohypophyses (Stefaneanu et al. 1994a). In nontumorous pituitaries, in situ hybridization combined with immunocytochemistry demonstrated estrogen receptor mRNA in adenohypophysial cells which were immunoreactive for growth hormone, or prolactin, or ACTH, or TSH, or FSH/LH. The hybridization signal was most intensive in prolactin immunoreactive cells. Estrogen receptor mRNA was also revealed in Crooke's cells, posterior lobe-corticotrophs, squamous nests of the pars tuberalis and epithelial cells lining the pars intermedia cavities. The posterior lobe, capillary endothelium and connective tissue expressed no estrogen receptor gene. Estrogen receptor mRNA was present in all adenoma types including somatotroph adenomas, lactotroph adenomas, mixed somatotroph-lactotroph adenomas, mammosomatotroph adenomas, acidophil stem cell adenomas, functioning and silent corticotroph adenomas, thyrotroph adenomas, gonadotroph adenomas, null cell adenomas and oncocytomas. The hybridization signal was strongest in lactotroph and mammosomatotroph adenomas. In lactotroph adenomas removed from patients treated with bromocriptine the hybridization signal was weak or absent suggesting that suppression of estrogen receptor gene plays a role in the inhibition of prolactin synthesis and tumor growth.

Dopamine receptors can also be localized using in situ hybridization methodology. This receptor which plays a fundamental role in the regulation of prolactin secretion could be demonstrated in all pituitary adenoma types (Stefaneanu et al. 1993). The hybridization signal was weak or moderate in tumors not exposed to dopamine agonists whereas it was very strong in lactotroph adenomas removed from patients treated with bromocriptine suggesting that dopamine agonists inhibit transcription of estrogen receptor gene and stimulate transcription of dopamine (D_2) receptor gene. These opposing effects on the levels of estrogen receptor and dopamine receptor mRNAs may play a pivotal role in the reduction of prolactin synthesis and release and in adenoma shrinkage.

These findings described briefly here conclusively show the value of in situ hybridization in the study of adenohypophysial cells and adenomas. We are convinced that in the near future more and more probes will be easily available and with further improvement in methodology, exciting novel results can be expected in the coming years.

Acknowledgement

This work was supported in part by the Medical Research Council of Canada, the St. Michael's Hospital Foundation, Mr. and Mrs. Jarislowsky and the Lloyd Carr-Harris Foundation. Author wishes to thank for all these awards. The contribution of Dr. Eva Horvath, Dr. Lucia Stefaneanu, and Dr. Kamal Thapar is gratefully acknowledged. Author is indebted to Ms. Elizabeth Chambers, Dr. Zi Cheng, Mrs. Anca Popescu and Mr. Fabio Rotondo for the participation in the studies and Ms. Lisa Horvath for the secretarial work.

References

1. Joubert (Bression) D, Benlot C, Lagoguey A, Garnier P, Brandi AM, Gautron JP, Legrand JC, Peillon F (1989) Normal and growth hormone (GH)-secreting adenomatous human pituitaries release somatostatin and GH-releasing hormone. J Clin Endocrinol Metab 68: 572–577
2. Kovacs K, Horvath (1986) Tumors of the pituitary gland. In : Atlas of tumor pathology. Series 2, Vol 21. Armed Forces Institute of Pathology, Washington
3. Levy A, Lightman SL (1992) Growth hormone-releasing hormone transcripts in human pituitary adenomas. J Clin Endocrinol Metab 74: 1474–1476
4. Stefaneanu L, Kovacs K, Horvath E (1993) In situ hybridization of dopamine D_2 receptor mRNA in human pituitary adenomas. Progr Abstr 75th Ann Meet, Endocr Soc (Abstr No 669), p 218
5. Stefaneanu L, Kovacs K, Horvath E, Lloyd RV, Buchfelder M, Fahlbusch R, Smyth H (1994a) In situ hybridization study of estrogen receptor messenger ribonucleic acid in human adenohypophysial cells and pituitary adenomas. J Clin Endocrinol Metab 78: 83–88
6. Stefaneanu L, Thapar K, Kovacs K, Horvath E, Lloyd RV (1994b) In situ hybridization study of growth hormone releasing hormone (GRH) mRNA in human pituitary adenomas. Progr Abstr 76th Ann Meet Endocr Soc, (Abstr No 782), p 396
7. Thapar K, Kovacs K, Laws ER Jr, Muller PJ (1993) Pituitary adenomas: current concepts in classification, histopathology, and molecular biology. Endocrinologist 3: 39–57
8. Wakabayashi I, Inokuchi K, Hasegawa O, Sugihara H, Minami S (1992) Expression of growth hormone (GH)-releasing factor gene in GH-producing pituitary adenoma. J Clin Endocrinol Metab 74: 357–361

Correspondence: Kalman Kovacs, M.D., Department of Pathology, St. Michael's Hospital, University of Toronto, 30 Bond Street, Toronto, Ontario M5B 1W8, Canada.

Acta Neurochir (1996) [Suppl] 65: 7–10

Molecular Biology of Growth-Hormone-Secreting Human Pituitary Tumours: Biochemical Consequences and Potential Clinical Significance

E.F. Adams, M. Buchfelder, T. Lei, B. Petersen, and **R. Fahlbusch**

Department of Neurosurgery, University of Erlangen-Nürnberg, Erlangen, Federal Republic of Germany

Summary

Molecular biological studies have revealed that 30–40% of GH-secreting human pituitary tumours, associated with acromegaly, harbour single-base missense mutations within the Gsα gene, termed gsp oncogenes. In addition, a large proportion of GH-secreting tumours inappropriately express the GH-releasing factor (GRF) gene. Gsp-oncogenes result in elevated adenylyl cyclase activity with consequent abnormally high cAMP production. In culture, GH-secreting tumours expressing gsp oncogenes respond more efficiently to the somatostatin analogue, octreotide (SMS), raising the possibility that acromegalics harbouring gsp-positive tumours may be those who optimally benefit from SMS therapy. Inappropriate expression of GRF may result in abnormal presence of a positive autocrine feedback loop, in which secreted GRF acts on the same cells to promote cellular proliferation and GH secretion. Blockade of GRF mRNA translation by means of anti-sense oligonucleotide approaches may prove to be of value in inhibiting tumour function.

Keywords: Acromegaly; molecular biology; gsp-oncogenes; GRF production.

Introduction

The finding that human pituitary tumours are monoclonal in origin indicates that they probably arise as a consequence of somatic mutations at the DNA level [3]. In recent years, the application of DNA technology has led to significant advances in the elucidation of the somatic defects which can occur in pituitary tumours, including oncogene expression [5,7], allele loss [4] and inappropriate gene transcription and translation [14]. One of the most exciting advances is the discovery that 30–40% of human growth hormone (GH)-secreting tumours carry somatic missense mutations within the gene for the α-subunit of the stimulatory GTP-binding protein, Gs (Gsα), which controls adenylyl cyclase activity [12]. These mutations are examples of oncogene expression leading to tumorigenesis. GH-secreting tumours may also exhibit inappropriate gene transcription and translation. In this system, GH-releasing factor (GRF), normally produced in the hypothalamus, is abnormally expressed by the tumorous pituitary GH cells themselves [14]. We herein review these two aspects of GH-secreting tumour molecular biology, and discuss their potential clinical significance based on results of some of our own experimental studies.

GSP Oncogenes

In pituitary GH cells, the Gs protein couples the cell-surface GRF receptor with intracellular adenylyl cyclase [8]. G proteins are heterotrimeric complexes, consisting of α, β and γ subunits. In the active state, Gα subunits bind guanosine 5'-triphosphate (GTP) and possess GTPase activity that converts the G protein back to the inactive form in which GDP is bound to the Gα subunit. Thus, for Gs proteins in GH cells, GRF receptor activation leads to dissociation of GDP and binding of GTP to the Gsα subunit which, in turn, results in activation of adenylyl cyclase and cAMP production. Regulation of the system is achieved by the intrinsic GTPase activity of the Gsα subunit, which hydrolyses the bound GTP to GDP thus promoting a return to the basal state. In 30–40% of GH-secreting human pituitary tumours, somatic mutations within the coding region of the Gsα gene result in amino acid substitutions which eliminate the intrinsic GTPase activity and thus ultimately lead to constitutive adenylyl cyclase (Fig. 1) [12]. The resultant elevated cAMP production is thought to be the cause of excessive GH secretion and GH cell proliferation in

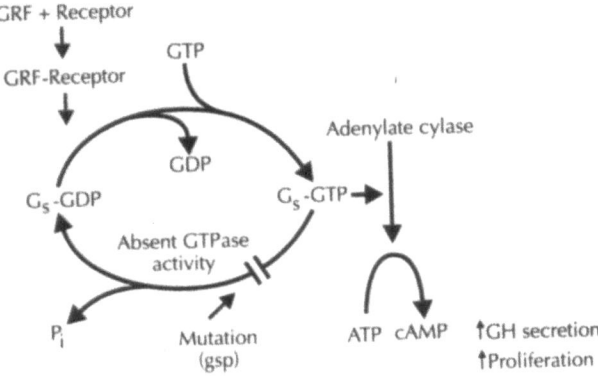

Fig. 1. Site of action of gsp-oncogenes in the GRF/GRF-receptor binding signal transduction cascade

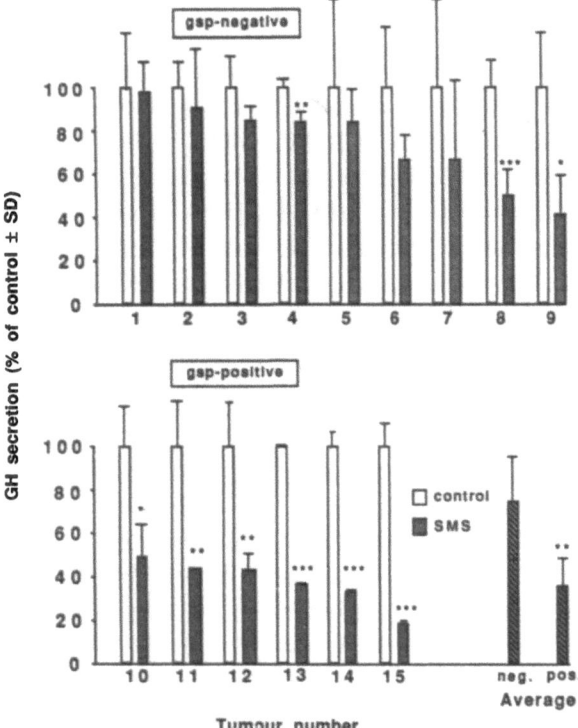

Fig. 2. In-vitro effect of SMS (10 nmol/L) on GH secretion by 9 gsp-negative (top panel) and 6 gsp-positive (lower panel) human pituitary tumours in cell culture. * p < 0.05, ** p < 0.01, *** p < 0.001 v. control

this sub-set of pituitary tumours. The mutations occur in either codon 201 or 227 [6,12]. The wild-type sequence in codon 201 is CGT (arginine) and this may mutate to TGT (cysteine) or CAT (histidine). In codon 227, the wild-type CAG (glutamine) sequence may mutate to CGG (arginine) or CTG (leucine). Mutated Gsα subunit genes have been termed gsp oncogenes [11].

GH-secreting tumours can therefore be grouped into those expressing gsp oncogenes (gsp-positive) and those without Gsα mutations (gsp-negative). Several studies have been directed at determining the clinical and biochemical characteristics of acromegalics harbouring gsp-positive pituitary tumours [1,2,9,13, 17,18]. Although there are conflicting data concerning tumour size, proliferative potential and GH secretory activity, most studies indicate that gsp-positive tumours retain higher responsiveness to GH-secretion inhibitory influences such as somatostatin, dopamine and an oral glucose load [1,13,17]. To further investigate this, we have examined the in vitro response of GH secretion to the somatostatin analogue, Octreotide (SMS), in groups of gsp-positive and gsp-negative pituitary tumours. Freshly resected human GH-secreting tumours, removed from acromegalic patients, were dispersed with collagenase and subjected to cell culture as previously described [2]. Cells were exposed for 4 hours to SMS (10 nmol/L) after which the amount of GH secreted was determined by an ELISA technique. Presence of gsp oncogenes was assessed by direct sequence analysis of appropriate regions of the Gsα gene generated by the polymerase chain reaction (PCR), as described [1]. Figure 2 summarises the findings. Of 15 tumours examined, 9 were gsp-negative (Fig. 2, top panel), and 6 (40%) were gsp-positive (Fig. 2, lower panel). Of the 9 gsp-negative tumours, SMS failed to significantly suppress GH secretion in 6. In the remaining 3 tumours, suppression of GH secretion was greater than 50% in only one. In marked contrast, GH secretion was significantly (p < 0.05) suppressed in all of the gsp-positive tumours, and this was greater than 50% in 5 of the 6 studied. These results are in agreement with the concept that gsp-positive tumours retain greater responsiveness to somatostatin. Furthermore, since preoperative treatment with SMS may have excellent clinical benefit in about 50% of acromegalics [15], our results raise the possibility that presence of gsp oncogenes can be used as markers for optimal responsiveness to SMS. Further studies are required to confirm this and we are currently investigating preoperative responses to SMS in acromegalics with and without gsp oncogenes in terms of effects on tumour size, tumour consistency and serum GH levels.

Inappropriate Expression of GRF

In normal humans, GH secretion is stimulated by GRF produced in the hypothalamus and transported to the pituitary via the hypothalamo-hypophysial portal

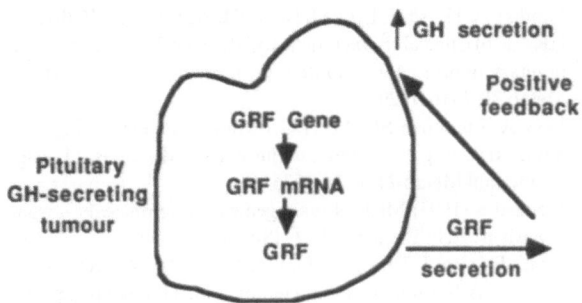

Fig. 3. Schematic representation of potential positive autocrine feedback mechanism in GH-secreting pituitary tumours abnormally expressing the GRF gene

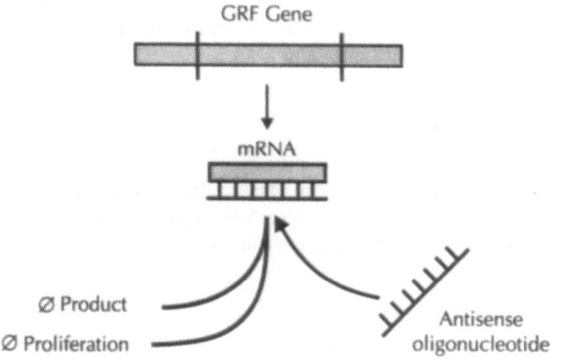

Fig. 4. Principle of anti-sense oligonucleotide therapy. Blockade of mRNA translation would result in elimination of the gene product and thus interuption of the autocrine feedback loop

blood system. Using molecular biological techniques and cell culture, considerable evidence has accumulated showing that tumorous GH pituitary cells themselves may abnormally produce and secrete GRF [10, 14,18]. For example, in situ hybridisation, in which frozen pituitary sections were incubated with radioactively labelled oligonucleotides against the second exon of the GRF gene, revealed cellular presence of GRF mRNA transcripts in 13 of 17 (76%) GH-secreting tumours, but not in normal pituitary cells [14]. Similar results were obtained by the technique of reverse-transcription-PCR, in which RNA was extracted from normal and tumorous pituitary tissues, DNA copies (cDNA) of the RNA made by use of the enzyme reverse transcriptase, and GRF mRNA sequences identified by PCR [18]. Complementary to these findings, active secretion of GRF peptide by GH pituitary cells has also been described [10]. Although the evidence is as yet circumstantial, these results suggest that an abnormal autocrine loop exists in at least some GH-secreting tumours, in which the GRF gene is inappropriately transcribed and translated to GRF peptide, the secretion of which exerts a feedback effect on

the same cells to elicit not only GH secretion but possibly also increased proliferation (Fig. 3). Support for this latter concept is provided by the observations that excessive ectopic GRF production is associated with pituitary GH cell hyperplasia [16]. Indeed, in our own preliminary experiments, we have demonstrated that the mouse ACTH-secreting pituitary cell line, AtT20, contains corticotrophin-releasing-factor (CRF) mRNA and that an anti-sense oligonucleotide against CRF mRNA is able to reduce thymidine uptake (Lei, Adams and Fahlbusch, unpublished). Such findings suggest novel approaches to the treatment of pituitary tumours, based on anti-sense technology. Since pituitary tumour development may be the result of inappropriate gene expression, blockade of translation by introduction of anti-sense sequences into the tumorous cells may results in decreased hormone secretion and cellular proliferation (Fig. 4). This remains an exciting area for future research.

Conclusions

Through the use of molecular biological techniques, important insights have been attained into the defects which can occur at the DNA level in GH-secreting tumours, including somatic point mutations and inappropriate gene expression, as exemplified by gsp oncogenes and GRF production. The biochemical consequences of these defects may have clinical implications in designing optimal treatment regimes and developing novel therapeutic approaches.

Acknowledgements

We are grateful to the Deutsche Forschungsgemeinschaft (ref: Ad 100/2-1) for financial support.

References

1. Adams EF, Brockmeier S, Friedman E, Roth M, Buchfelder M, Fahlbusch R (1993) Clinical and biochemical characteristics of acromegalic patients harboring gsp-positive and gsp-negative pituitary tumors. Neurosurgery 33: 198–203
2. Adams EF, Buchfelder M, Petersen B, Fahlbusch R (1994) Effect of pituitary adenylate-cyclase activating polypeptide on human somatotrophic tumours in cell culture. Endocrine 2: 75–79
3. Alexander JM, Biller BMK, Bikkal H, Zervas NT, Arnold A, Klibanski A (1990) Clinically non-functioning pituitary tumours are monoclonal in origin. J Clin Invest 86: 336–340
4. Boggild MD, Jenkinson S, Pistorello M, Boscaro M, Scanarini M, McTernan P, Perrett CW, Thakker RV, Clayton RN (1994) Molecular genetic studies of sporadic pituitary tumors. J Clin Endocrinol Metab 78: 387–392

5. Bourne HR (1987) Discovery of a new oncogene in pituitary tumours? Nature 330: 517–518

6. Clementi E, Malgaretti N, Meldolesi J, Taramelli R (1990) A new constitutively activating mutation of the Gs protein α subunit-gsp oncogene is found in human pituitary tumours. Oncogene 5: 1059–1061

7. Daniels M, Harrison D, Raghaven R, Birch P, Ince PG, James RA, Kendall-Taylor P (1993) C-myc immunoreactivity in human pituitary tumours. J Endocrinol 137 [Suppl]: 147

8. Faglia G (1993) Epidemiology and pathogenesis of pituitary adenomas. Acta Endocrinol 129 [Suppl 1]: 1–5

9. Harris PE, Alexander JM, Bikkal HA, Hsu DW, Hedley-Whyte ET, Klibanski A, Jameson JL (1992) Glycoprotein hormone α-subunit production in somatotroph adenomas with and without Gsα mutations. J Clin Endocrinol Metab 75: 918–923

10. Joubert D, Benlot C, Lagoguey A, Garnier P, Brandi AM, Gautron JP, Legrand JC, Peillon F (1989) Normal and growth hormone (GH)-secreting adenomatous human pituitaries release somatostatin and GH-releasing hormone. J Clin Endocrinol Metab 68: 572–577

11. Klibanski A (1990) Editorial: further evidence for a somatic mutation theory in the pathogenesis of human pituitary tumors. J Clin Endocrinol Metab 71: 1415A–1415C

12. Landis CA, Masters SB, Spada A, Pace AM, Bourne HR, Vallar L (1989) GTPase inhibiting mutations activate the α chain of Gs and stimulate adenylyl cyclase in human pituitary tumours. Nature 340: 692–696

13. Landis CA, Harsh G, Lyons J, Davis RL, McCormick F, Bourne HR (1990) Clinical characteristics of acromegalic patients whose pituitary tumors contain mutant Gs protein. J Clin Endocrinol Metab 71: 1416–1420

14. Levy A, Lightman SL (1992) Growth hormone-releasing hormone transcripts in human pituitary adenomas. J Clin Endocrinol Metab 74: 1474–1476

15. Melmed S (1993) Medical management of acromegaly – what and when? Acta Endocrinol 129 [Suppl 1]: 13–17

16. Rivier J, Spiess J, Thorner MO, Vale W (1982) Characterization of a growth hormone-releasing factor from a human pancreatic islet tumor. Nature 300: 276–278

17. Spada A, Arosio M, Bochicchio D, Bazzoni N, Vallar L, Bassetti M, Faglia G (1990) Clinical, biochemical, and morphological correlates in patients bearing growth hormone-secreting pituitary tumors with and without constitutively active adenylyl cyclase. J Clin Endocrinol Metab 71: 1421–1426

18. Wakabayashi I, Inokuchi K, Hasegawa O, Sugihara H, Minami S (1992) Expression of growth hormone (GH)-releasing factor gene in GH-producing pituitary adenoma. J Clin Endocrinol Metab 74: 357–361

Correspondence: E.F. Adams, Ph.D., Department of Neurosurgery, University of Erlangen-Nürnberg. Schwabachanlage 6, D-91054 Erlangen, Federal Republic of Germany.

Acta Neurochir (1996) [Suppl] 65: 11–12

Surgical Results in Microadenomas

M. Giovanelli, M. Losa, P. Mortini, S. Acerno, and **E. Giugni**

Department of Neurosurgery, San Raffaele IRCCS, University of Milano, Milano, Italy

Summary

Pituitary microadenomas are small tumors whose maximal diameter is less than 1 cm. The aim of surgical removal of microadenomas should be not only the reversal of hormone hypersecretion but also the preservation of normal anterior pituitary function. Our series includes 230 patients with a microadenoma who had their first operation in our department: 45 were GH-secreting, 92 were PRL-secreting, 90 were ACTH-secreting, and 3 were TSH-secreting. Remission of disease was achieved in 81%, 77%, 91%, and 100% of GH-, PRL-, ACTH-, and TSH-secreting adenomas, respectively. There was no perioperative mortality and only 5 patients experienced a major complication. A total of 7 patients had diabetes insipidus for at least 6 months after operation. Hypopituitarism, not present in any patients before operation, developed in 3.5% of the cases. Our experience confirms that patients with microadenomas have the best chances of a successful operation. Since tumor size should gradually increase with time, we underscore the need of early diagnosis and treatment in patients with pituitary adenomas.

Keywords: Pituitary neoplasms; transsphenoidal surgery; recurrence; complications.

Introduction

Harvey Cushing, the pioneer of pituitary surgery, already in 1927 mentioned the theoretical possibility to avoid the long-term devastating effect of the small hypersecreting pituitary adenomas by means of surgical removal. About 40 years later, following the introduction of intraoperative fluoroscopy and the use of the operative microscope, Jules Hardy reported the first operative results in hypersecreting microadenomas. At that time there still was skepticism about the possibility to cure the pituitary microadenomas by surgery alone. However, as the experience of Hardy was replicated by other neurosurgeons, transsphenoidal surgery became the first choice treatment of most hypersecreting microadenomas. In the present study we review our experience in the surgical treatment of pituitary microadenomas with special regard to the comparison with all the other pituitary macroadenomas.

Material and Methods

Between 1970 and 1993, 839 patients received their first operation at the University of Milan for a pituitary adenoma. Of these 230 (27.4%) had a microadenoma (maximum diameter ≤ 10 mm). The tumors were classified according to the type of hormone hypersecretion: 45 were GH-secreting (18.4% of all the GH-secreting tumors), 92 were PRL-secreting (43.8%), 90 were ACTH-secreting (88.2%), and 3 were TSH-secreting (20.0%). In the series there was no nonfunctioning tumor, reflecting our policy not to operate small pituitary lesions in the absence of hormonal derangements. Among all the 230 operated microadenomas, PRL- and ACTH-secreting tumors were the most frequent (40.0% and 39.1%, respectively) followed by GH-secreting (19.6%) and TSH-secreting adenomas (1.3%).

Results

There was no surgical mortality in this series of 230 patients with microadenomas, whereas major morbidity occurred in 5 patients: 1 case of carotid artery injury, 2 cases of postoperative seizures (7 and 9 days after operation), and 2 cases of CSF leak requiring surgical repair. In contrast, there were 7 deaths in patients operated for a macroadenoma (1.1%) and major morbidity occurred in 22 patients (3.6%).

Normalization of GH and/or insulin growth factor-1 (IGF-1) levels was achieved in 37 of the 45 acromegalic patients with a microadenoma (81.8%), as compared to only 93 of the other 200 patients with a macroadenoma (46.4%). Preoperative basal GH levels and invasiveness of the tumor retained their prognostic value also in the subgroup of patients with GH-secreting microadenomas. In fact, choosing an arbitrary cut-off level of $40\,\mu g/l$, only 4 of the 37 patients cured after surgery (10.8%) had a basal preoperative GH value higher, as compared to 4 of the 8 not cured (50.0%). Similarly, among cured patients, only 3 (8.1%) showed signs of tumor invasiveness, as compared to 5 of the 8 not cured patients (62.5%).

Normalization of PRL secretion occurred in 71 of the 92 patients operated for a microprolactinoma (77.2%). This rate of cure was clearly better than that obtained in patients with macroprolactinomas (26 out of 118 patients; 22.0%). Since there is still debate about the possible negative influence of preoperative therapy with dopaminergic drugs on the rate of success of subsequent surgery, we analyzed our results according to this variable. Sixty-seven patients (72.8%) had received previous therapy with dopaminergic drugs for at least 4 months before surgery. Mean age and basal PRL levels did not differ significantly between the two groups. The success rate of surgery in previously treated patients (74.6%) was slightly but nonsignificantly lower than that in patients who never received dopaminergic therapy (84.0%).

Patients with Cushing's disease were considered in remission after surgery if they had resolution of symptoms and signs of hypercortisolism, low or normal ACTH and cortisol levels, and normal suppressibility of cortisol levels during the low-dose dexamethasone test. By these criteria, 80 of the 90 (88.9%) patients with an ACTH-secreting microadenoma were cured. A similar success rate was achieved also in patients with macroadenomas (10 out of 12 patients; 83.3%). The lack of a negative influence of tumor size on the outcome of surgery is easily accounted for by the fact that most ACTH-secreting macroadenomas (9 of 12) were actually intrasellar and smaller than 15 mm.

Resolution of hyperthyroidism with normalization of free thyroid hormone and of α-subunit levels, when increased preoperatively, occurred in all the 3 patients with a TSH-secreting microadenoma. A similar good result was observed in all the 4 patients with an intrasellar macroadenoma, whereas the rate of cure declined to 62.5% in the 8 patients with an extrasellar extending macroadenoma.

In summary, as a whole 191 of the 230 patients (83.0%) operated for a pituitary microadenoma were considered cured after surgery. The outcome of surgery in the 4 types of secreting tumors is reported in Fig. 1.

Normal pituitary function was preserved in most patients. Postoperative diabetes insipidus, defined as polyuria lasting for at least 6 months, occurred in 7 patients (3.0%). Only patients with Cushing's disease had a relatively high frequency of this complication

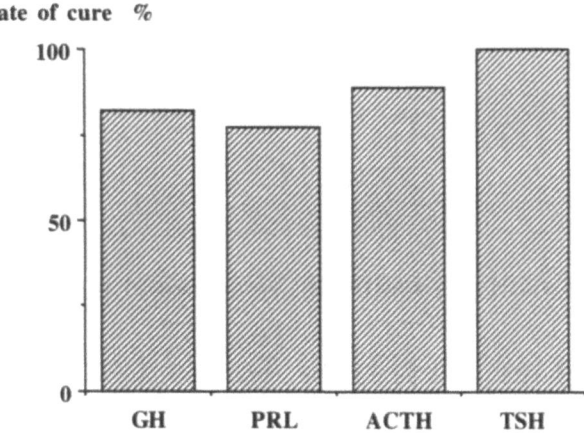

Fig. 1. Overall rate of surgical cure in pituitary microadenomas subdivided according to the type of hormone hypersecretion

(6.7%), probably due to the more frequent growth of the ACTH-secreting tumors near or inside the neurohypophysis. The onset of new partial or total hypopituitarism after surgery occurred in 8 of the 230 patients (3.5%); however, in only 3 cases hypopituitarism was complete.

Recurrence of the disease developed in 16 of the 191 successfully operated patients (8.4%). The probability of the disease to recur was seen, in decreasing order of frequency, in patients with Cushing's disease (10 of 80 patients; 12.5%), microprolactinomas (5 of 71 patients; 7.0%), and acromegaly (1 of 37 patients; 2.7%).

Conclusions

In our experience, microadenomas represent about one fourth of all operated pituitary adenomas. Operative mortality in our series was 0% and major morbidity was low (less than 3%), underscoring the safety of the transsphenoidal approach. Surgical removal of the pituitary adenoma was also very effective, leading to immediate remission of the disease in more than 80% of the patients, with no major difference in respect to the type of the adenoma. However, the long-term results were less favorable because of the occurrence of relapses of the pituitary adenomas in about 8% of the successfully operated patients. When compared to the results in large tumors, patients with microadenomas showed far better results both in terms of efficacy and safety. Our results strongly support a policy of early diagnosis of these rare tumors in order to improve the probability of cure following surgical therapy.

Correspondence: M. Giovanelli, M.D., Department of Neurosurgery, University of Milano, 1-20132 Milano, Italy.

Acta Neurochir (1996) [Suppl] 65: 13–15

Management of Huge Pituitary Adenomas

K. Takakura[1] and **A. Teramoto**[2]

[1]Department of Neurosurgery, Tokyo Women's Medical College and [2]Department of Neurosurgery, Nihon Medical College, Japan

Summary

Management of huge pituitary adenoma (more than 5 cm in diameter) is one of the most important issues on the treatment of pituitary tumors. We have analyzed the therapeutic modality and the result of our cases. From 1967 to 1983, 50 patients with huge adenoma (14.1%) out of a total 354 pituitary adenomas were surgically treated. The operative mortality was 25% for radical transcranial (TC) approach (10/40), 14% (1/7) for transsphenoidal (TS) approach and 0% (0/3) for combined two stage operations. From long-term follow-up, excellent prognoses were observed in only 44% of the patients treated by radical TC operation. After 1984, we have employed partial removal by TS surgery at the first stage, followed by reoperation by TS or TC surgery with or without radiotherapy or bromocriptine in case by case. Seventeen huge pituitary adenomas out of a total 700 pituitary adenomas were operated. There was no mortality nor major complications. The two stage operation with initial TS surgery is recommended for the management of huge pituitary adenomas.

Keywords: Pituitary adenoma; huge adenoma; prolactinoma; non-functioning adenoma; gonadotropinoma.

Introduction

With the recent advances of surgical techniques, the operative results of pituitary adenoma have markedly improved, accompanied by low mortality and equally low morbidity. Zervas [1] reported in 1984 that the operative mortality was 0.27% (7/2,606) in microadenomas and 0.9% (23/2,677) in macroadenomas from the international survey on results of transsphenoidal surgery. However, the surgical outcomes of large pituitary adenomas were sometimes not satisfactory [2–7], especially in cases of the radical removal.

In this paper, firstly, we analyzed our previous surgical results of huge pituitary adenomas treated before 1983, which were defined as adenomas larger than 5 cm in diameter or those extending over the level of Monro's foramen. Based on these data, we have developed the new management for each type of huge pituitary adenomas. Our recent results for these adenomas

are excellent without any mortality and morbidity as shown in the later part of this paper.

Previous Surgical Results of Huge Pituitary Adenomas

A total of 354 patients with pituitary adenomas underwent surgery in Tokyo University Hospital from 1967 to 1983. Among them, 50 patients (14.1%) had huge pituitary adenomas. They were 34 males and 16 females. The age and sex distribution are shown in Fig. 1. Endocrinological functions were elucidated in 35 adenomas, which consisted of 21 non-functioning, 12 prolactin-secreting and two GH-secreting adenomas.

Huge adenomas were resected by transcranial approach in 40 patients, by transsphenoidal approach in 7 patients and by combined surgeries in 3 patients. Eleven patients (22%) died within one month after

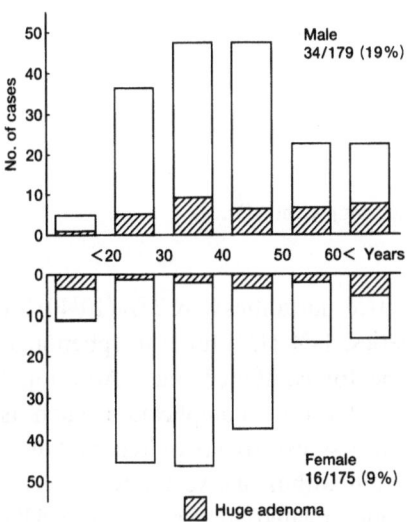

Fig. 1. Age and sex distribution of huge pituitary adenomas

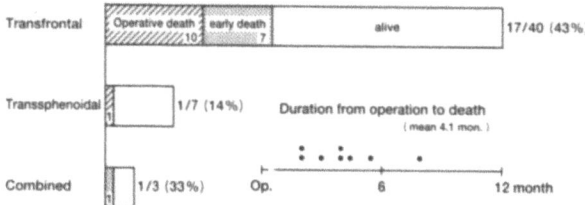

Fig. 2. Surgical approaches and early postoperative death

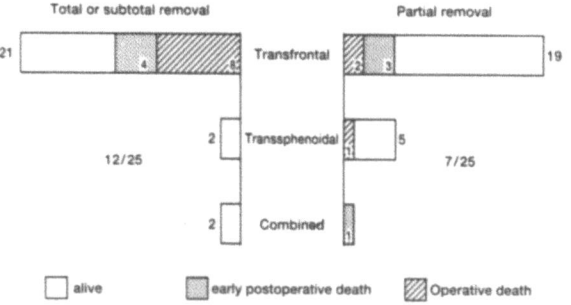

Fig. 3. Extent of tumor and early postoperative death

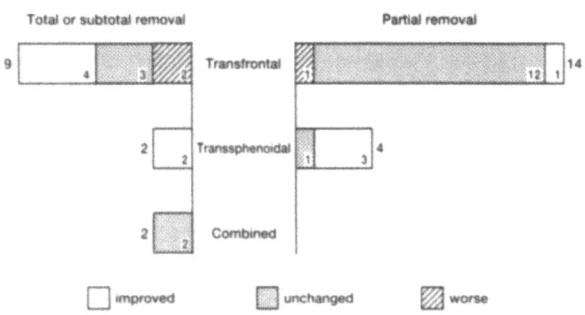

Fig. 4. Operative results for visual failures

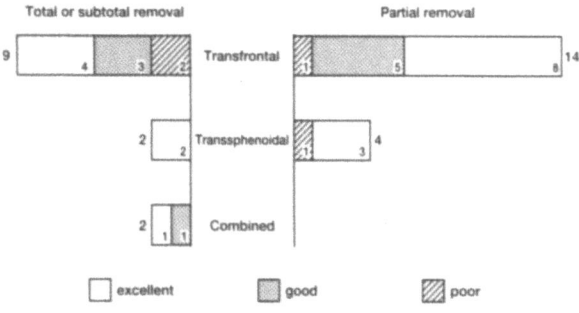

Fig. 5. Operative results for present ADL

surgery. The operative mortalities were 25% (10/40) for transcranial surgeries, 14% (1/7) for transsphenoidal surgeries and none for combined ones. Additional eight patients died of various complications such as prolonged consciousness disturbance or hypothalamo-pituitary malfunction within one year after surgery. Thus the death rates related to surgery were 43% (17/40) for transcranial surgery, 14% (1/7) for transsphenoidal surgery and 33% (1/3) for combined ones (Fig. 2).

Figure 3 shows the relation between postoperative deaths and surgical radicality in each approach. The highest death rate (57%) was seen in the group of radical removal by transcranial surgery. The rate decreased to 26% in the group of partial removal by transcranial surgery. One patient died in each partially removed group of transsphenoidal surgery and of combined ones.

Postoperative prognoses of visual failures could be evaluated in 31 patients (Fig. 4). Eighteen patients (58%) showed no significant changes before and after surgery, whereas improvements or aggravations were found in 11 (35%) or 3 (10%) patients respectively. Although most patients (12/14) of partially removed group showed no changes after transcranial surgery, the radical removal did not always lead to the visual improvements. On the contrary, the results of transsphenoidal surgery were much better regardless of surgical radicality (improved, 5/6).

Long-term functional prognoses could be obtained in 31 patients (Fig. 5). Of them, 18 patients (58%) were classified as excellent, 9 patients (29%) as good and 4 patients (13%) as poor. After transcranial surgery, excellent prognoses were found in 44% of patients with radical removal and in 57% of those with partial removal. On the other hand, 83% of patients presented an excellent prognosis after transsphenoidal surgery.

In summary, the radical tumor removal by transcranial surgery led to the highest visual problems and also long-term poor prognoses. Transsphenoidal and/or combined surgery were much safer and led to better results in patients with huge pituitary adenomas.

Recent Management of Huge Pituitary Adenomas

Based on this background, we always adopt transsphenoidal approach as the first surgery for huge pituitary adenomas, although adjunctive therapies may be sometimes necessary before operation. Our present management for several kinds of adenomas will be presented below.

Bromocriptine is rarely effective for so-called non-functioning adenomas, although a few cases were reported to show a minimal size reduction with this treatment. Therefore, transsphenoidal surgery should be firstly selected in order to reduce the main mass. Usually, we remove the whole intrasellar tumor and a part of suprasellar mass. We do not resect the additional tumor unless the suprasellar mass descends spontaneously into the sellar space. We think that the theoretical plane of diaphragma sellae may be the upper limit for the safe transsphenoidal surgery. We don't use the intrathecal injection of air or saline from the spinal catheter, whereas Valsalva's maneuver is always tried not only for pressing down the tumor but also for the confirmation of hemostasis. Too much intrasellar packing should be avoided for the first surgery in order to induce the descent of the residual suprasellar tumor. We use a small amount of oxidized cellulose and fibrin glue for the packing and usually do not repair the sellar floor, unless CSF leakage occurs during surgery.

Then we follow the patients for a couple of months using CT scan or MRI every two weeks. In most cases, the suprasellar portion, more or less, is going down into the vacant sellar space within three months. The second transsphenoidal surgery should be tried before the mucosal adhesion is complete (within three months). Although the basic principle of the second surgery is similar to that of the first one, the sellar packing and the repair of the floor must be done after removal of the tumor.

The next step of the treatment depends on the volume and location of the residual tumor. If a considerable amount of tumor still remains, we will remove it by transcranial approach followed by irradiation. Small amount of residual tumor is treated with conventional irradiation or gamma knife radiosurgery.

For huge prolactinomas, bromocriptine is usually administered for two or three months before surgery. The tumor size reduces significantly toward the sella turcica except for a few bromocriptine-resistant tumors. The tumor does not show a fibrous change with this relatively short-term treatment. Then, we remove the tumor of the whole intrasellar portion by transsphenoidal surgery. As mentioned above, we do not extend the operative maneuver unless the tumor of suprasellar portion comes down spontaneously into the sellar space. However, in the sellar packing the repair of the floor is necessary for the surgery of prolactinomas, because bromocriptine sometimes induces CSF rhinorrhea by reducing the size of invasive tumor. Postoperatively, most patients showed the rapid improvement of visual failures or extra-ocular muscle palsies. Then, we resume the administration of bromocriptine, the dose of which ranges from 7.5 to 22.5 mg/day according to the sensitivity to this treatment. As a rule, radiation therapy is not indicated for prolactinomas except for bromocriptine resistant ones.

A huge adenoma is relatively rare in acromegalic patients. Bromocriptine is sometimes used as the preoperative therapy of GH secreting adenomas, though it does not induce the marked size reduction in most cases except for some mixed GH cell-PRL cell adenomas. Octreotide (somatostatin analogue) is occasionally administered before surgery, whereas it is not so effective for tumor shrinkage. It was reported that the size reduction (20–30% shrinkage) occurred in 10–50% or patients treated with octreotide [8]. Since octreotide must be given as the injectile form (twice or three times a day), it cannot be used routinely as the adjunct therapy before surgery. If the tumor is resistant to bromocriptine and/or octreotide, the operative policy is essentially similar to that of non-functioning adenomas. After surgery, most patients are treated with irradiation as well as bromocriptine.

According to the above strategies, we have treated 17 patients with huge pituitary adenomas among 700 surgically treated cases in recent ten years. They included 12 non-functioning, 4 PRL producing and one GH producing adenomas. Neither operative mortality nor major complications were observed in these patients.

References

1. Zervas NT (1984) Surgical results for pituitary adenomas: results of an international survey. In: Black P McL *et al* (eds) Progress in endocrine research and therapy, Vol 1. Raven, New York, pp 377–385
2. Jefferson G (1940) Extrasellar extensions of pituitary adenomas. President's address. Proceedings of the Royal Society of Medicine 33: 433–458
3. Bakay L (1950) The results of 300 pituitary adenoma operations (Professor Herbert Olivecrona's series). J Neurosurg 7: 240–255
4. Jefferson A (1969) Chromophobe pituitary adenomata: the size of the suprasellar portion in relation to the surgery of operation (abstract). J Neurol Neurosurg Psychiatry 32: 633
5. Wirth P, Schwartz HG, Schwestschenau PR (1974) Pituitary adenomas: factors in treatment. Clin Neurosurg 21: 8–25
6. Symon L, Jakubowski J, Kendall B (1979) Surgical treatment of giant pituitary adenomas. J Neurol Neurosurg Psychiatry 42: 973–982
7. Symon L, Jakubowski J (1979) Transcranial management of pituitary tumours with suprasellar extension. J Neurol Neurosurg Psychiatry 42: 123–133
8. Lambert SW (1988) The role of somatostatin in the regulation of anterior pituitary hormone secretion and the use of its analogs in the treatment of human pituitary tumors. Endocr Rev 9: 417–438

Correspondence: Kintomo Takakura, M.D., Department of Neurosurgery, Tokyo Women's Medical College, 8-1, Kawadacho, Shinjukuku, Tokyo 162, Japan.

Acta Neurochir (1996) [Suppl] 65: 16–17

The Role of Transcranial Surgery in the Management of Pituitary Adenoma

R. H. Patterson

Division of Neurosurgery, The New York Hospital, Cornell Medical Center, New York, NY, U.S.A.

Summary

Transcranial pituitary surgery has a small, but distinct, role in the management of pituitary adenomas. Properly performed, a transcranial procedure should have an associated morbidity and mortality not much different than that of a transsphenoidal operation.

Keywords: Pituitary adenoma; surgery; transcranial; transsphenoidal; results.

Most surgery for pituitary adenoma is performed by a transnasal, transsphenoidal approach. The results are excellent, and the procedure is safe and well tolerated by patients. However, there are occasions where a transnasal approach is inadequate. As a consequence, transcranial pituitary surgery still has a place in the surgical tool chest. We recently reviewed the last 300 consecutive pituitary operations and found that 6 per cent, or 18, of them were performed by the transcranial route.

The reasons for the transcranial operation in our 18 cases are outlined in Table 1. In six cases, the tumor was giant or lobulated, in five it was too firm and fibrous to be dealt with satisfactorily through the nose, in five, the preoperative diagnosis was in doubt, one patient was a small child, and in another, removal of tumor from the cavernous sinus was planned. Interestingly, six of the eighteen had undergone a prior transsphenoidal operation.

As we have pointed out earlier, almost all firm tumors that are difficult to remove transsphenoidally are isointense on the T2 weighted image of the MRI [1]. It is therefore wise to warn patients whose MRI image has this characteristic that a second, transcranial operation may be required. We have found that 70 per cent of the T2 isointense tumors are firm but 30% are soft enough to allow an adequate transsphenoidal

removal. Prior radiation therapy is also a predictor of a firm tumor (Table 2).

Another indication for a transcranial approach is a tumor whose geometry will not allow a decent removal from below. This can be a lobulated tumor which is not located directly above the sella or a tumor that is dumbbell in shape. The dumbbell shape usually results from the herniation of tumor through an incompetent diaphragm sellae with further growth of tumor above the diaphragm. Normally, the diaphragm acts as the capsule of the tumor and slowly stretches as the tumor grows. If the diaphragm has a large opening for the pituitary stalk, then some of the tumor may grow through the hole and assume a dumbbell shape. The problem is getting the tumor above the diaphragm. There are various surgical tricks that may make this possible from below, but they do not always succeed, and a second operation from above may be necessary.

The results of transcranial surgery should be about the same as those of transsphenoidal surgery. We compared some of our results with vision in large tumors

Table 1. *Reasons for the Transcranial Operation*

Giant or lobulated	6
Fibrous	5
Uncertain diagnosis	5
Child	1
Cavernous sinus removal	1
(Six had previous transsphenoidal operation)	

Table 2. *Prior Treatment and Consistency of Pituitary Adenoma*

	No radiation	Prior radiation
Fibrous	6	4
Soft	74	3

Table 3. *Surgical Results in Patients with Impaired Vision*

	N	Normal	Improved	Unchanged	Worse
Transsphenoidal	70	17%	69%	14%	3%
Transcranial	94	45%	36%	16%	2%

Table 4. *Day Mortality in Transcranial Surgery for Pituitary Adenoma*

	N	Dead
First operation	138	0
Recurrence	27	2
	165	1.2%

with published results from the Mayo Clinic, and they were about the same with vision improving in 80 percent, staying the same in 18 per cent, and worsening in 2 percent (Table 3). In an earlier published series, we had no mortality in 138 primary transcranial operations and two deaths in 27 operations for recurrence, both in desperately ill patients [2] (Table 4).

Patients with colossal tumors pose the most difficult problem. If the tumor is a prolactinoma, then the administration of bromocriptine may shrink the tumor dramatically, sometimes even enough to result in a cerebrospinal fluid fistula. If the tumor is not a prolactinoma, then transcranial surgery is appropriate. Because the tumor is large and often lobulated, complete removal may be impossible. The problems come following surgery because there is a noticeable risk of postoperative hemorrhage in the bed of the tumor. This may come about as bleeding from residual tumor pure and simple, or it may result from a hemorrhagic infarction as surgery deprives some lobule of tumor of its blood supply.

The surgical strategies that can minimize the morbidity of transcranial pituitary are well known but worth emphasizing. These include positioning the patient with the head back 45 degrees so that the frontal lobe falls back from the floor of the frontal fossa, making the craniotomy opening low on the roof of the orbit, using spinal drainage, opening the arachnoid membranes to allow further brain relaxation, minimizing retraction, and avoiding all manipulation of stretched optic nerves.

References

1. Ray S, Patterson RH Jr (1962) Surgical treatment of pituitary adenomas. J Neurosurg 18: 1–8
2. Snow RB, Johnson CE, Morgello S, Lavyne MH, Patterson RH Jr (1990) Is magnetic resonance imaging useful in guiding the operative approach to large pituitary tumors? Neurosurgery 26: 801–803

Correspondence: Russel H. Patterson, M.D., Division of Neurosurgery, The New York Hospital, Cornell Medical Center, 525 East, 68th Street, New York, NY 10071, U.S.A.

Acta Neurochir (1996) [Suppl] 65: 18–21

Proliferation Parameters for Pituitary Adenomas

M. Buchfelder[1], R. Fahlbusch[1], E.F. Adams[1], F. Kiesewetter[2], and P. Thierauf[3]

[1]Neurochirurgische Klinik, Poliklinik, [2]Dermatologische Klinik, and [3]Pathologisches Institut, Universität Erlangen-Nürnberg, Federal Republic of Germany

Summary

In this review, the value of assessing proliferation parameters in surgically resected pituitary tumour tissue is analyzed. Histological examination of basal dura biopsies identifies invasive growth even when intraoperatively not apparent to the surgeon. Determination of DNA-polymerase activity, Ki-67 immunohistochemistry and DNA-flow-cytometry shows a clear difference in the proliferative potential of enclosed and invasive pituitary adenomas. Among the various endocrinologically differentiated groups ACTH-secreting adenomas associated with Nelson's syndrome and thyrotropinomas were the most rapidly proliferating. At present, however, our results reveal that the prognosis of an individual patient cannot be reliably predicted on the basis of such studies.

Keywords: Invasion; pituitary adenomas; immunohistochemistry; flow-cytometry; proliferation; transsphenoidal surgery.

Introduction

Pituitary adenomas arise from anterior pituitary cells and usually displace rather than invade the surrounding anatomical structures. Although generally considered to be slowly growing [9,11,12,16], clinical experience shows that a small percentage of these tumours grow aggressively. As early as 1952, Jefferson [9] published findings on an autopsy series of 13 adenomas associated with an unfavourable clinical outcome in which he documents severe invasion of cavernous sinus, brain and cranial base. He also pioneered the histological investigation of these invasive cases. However, only a few cases of pituitary carcinoma, in which distant metastases were documented [11], are reported in the literature. Efforts to study proliferation of these tumours aimed to differentiate the rather benign and the more aggressively growing tumours on the basis of investigation of materials which were obtained during surgery. This is a short review of methods that have been employed and their results.

Invasion

Pituitary adenomas are generally considered invasive if they have infiltrated or perforated the normal anatomical confines of the pituitary gland, namely the sellar diaphragm, the basal dura or the cavernous sinus. There is a considerable discrepancy in the literature concerning invasion recognized by the surgeon and histologically documented invasion. Up to 85% of pituitary adenomas show some signs of histological invasion if a subtle examination of the surrounding structures is performed [16,17]. Shaffi and Wrightson [18] proposed to biopsy arachnoid and basal dura. Scheithauer *et al.* [16] and Selman *et al.* [17] have analyzed a large surgical series with regard to functional classification. In surgical series, however, there is much less invasion recognized. When our surgical series is analyzed, which consists of 1670 operations performed between December 1982 and February 1994, earlier findings are basically confirmed (Fig. 1). Among them were 1540 transsphenoidal and 130 transcranial operations. During transsphenoidal operation we found a total rate of invasion of 22.3% of the cases, whereas 53.8% of pituitary adenomas operated transcranially were invasive. The apparently higher rate of invasion found in the cases dealt with by transcranial surgery is merely a reflection of the selection criteria for the different approaches. Among the different endocrinological diseases, invasion was most commonly found in Nelson's syndrome and thyrotropinomas, whereas in Cushing's disease it was found exceedingly rarely.

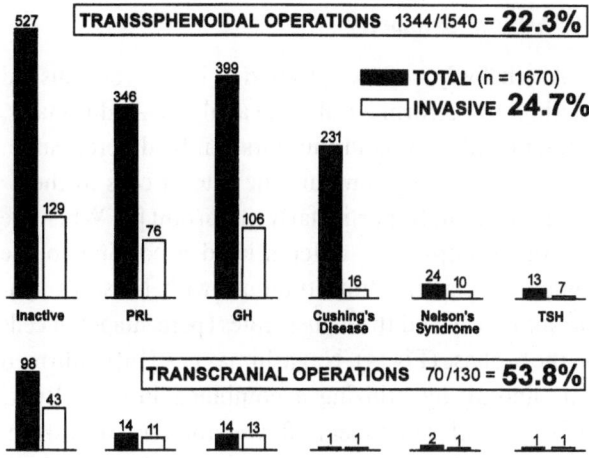

Fig. 1. Surgical invasion in the Erlangen surgical series of pituitary adenomas (1670 cases operated on between 1.12.82 and 28.02.94)

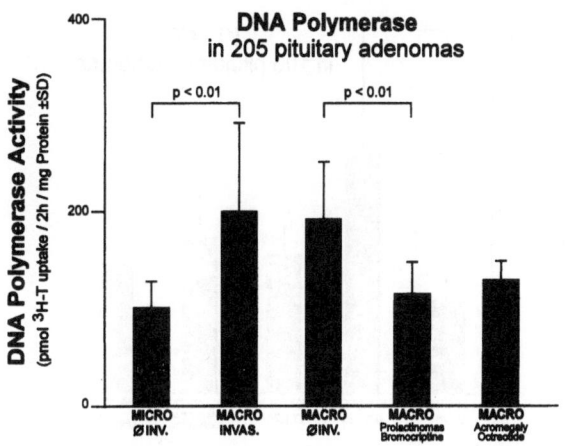

Fig. 2. DNA polymerase activity in 205 pituitary adenomas. The adenomas are grouped according to their size and invasive character

In order to discern how reliable the surgical impression really is, biopsies of the basal dura (endosteum) were performed. The endosteum is the most easily removable anatomical structure surrounding the adenoma. A comparison of recognition rates of surgical (n=164) and histological (n=323) invasion in 676 patients, in whom basal dura biopsies were obtained shows that the surgeon had only recognized about half of the cases as invasive. This suggests that in another half actually existing invasion is hidden for the surgeon. A more differentiated analysis shows that hidden invasion is more common in microadenomas (65%) than in macroadenomas (47%). Selman et al. [17] agree with this observation. We consider a penetration of the endosteum with tumour infiltration into the skull base or sphenoid sinus mucosa to be "massive invasion". The latter was most frequently found in thyrotropinomas and ACTH-secreting adenomas presenting with Nelson's syndrome, which both are commonly referred to as "feedback-tumours".

Assaying for DNA-Polymerase

Clinical observations and histological observations have failed to show whether there is an association between invasive and aggressive growth with increased proliferative potential. This relation was examined by utilizing various proliferation parameters in tissues resected during surgery. In order to study various degrees of malignancy in breast cancer, Edwards et al. [5] developed a biochemical assay for α-DNA polymerase, an enzyme with a pivot role in mitosis. Wiklund and Gorski [20] first employed a modified technique to study pituitary tumours. Utilizing a simi-

lar approach [2] we could show that there is a significant difference in the mean DNA-polymerase activity when large numbers of non-invasive microadenomas and invasive macroadenomas were compared (Fig. 2). A medical pretreatment by dopamine-agonists in macroprolactinomas and by somatostatin analogs in macroadenomas associated with acromegaly seemed to reduce the proliferative potential. Melmed et al. [13] showed a reduced DNA-polymerase activity in bromocriptine-treated as compared to untreated prolactinomas in rats. When endocrine tumour types were compared, adenomas associated with Nelson's syndrome and thyrotropinomas were shown to be the most rapidly proliferating tumours. In contrast to these groups which could clearly be differentiated, DNA polymerase activity measurements failed to reveal differences between tumour groups that were characterized by molecular biological investigations, e.g. mutations in the membrane-attached G_s-protein.

Immunohistochemistry for Proliferation-Associated Antigens

When Gerdes et al. [7] characterized a proliferation-associated nucleic antigen which proved to be detectable in all phases of the cell cycle except for the G_0- and early G_1-phase, an immunohistochemical method was available to measure proliferation of slowly-growing tumours in frozen section biopsies. It has been suggested that this technique is a more reliable measure of solid tumor proliferation than tritiated thymidine labeling. Using this technique, we could again show that there is a significant difference in the mean percentage of Ki-67 positive cells when microadenomas

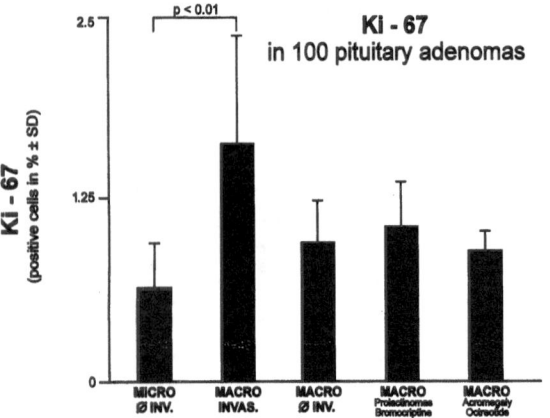

Fig. 3. Ki-67 immonohistochemistry in frozen sections of 100 pituitary adenomas. The adenomas are grouped according to their size and invasive character

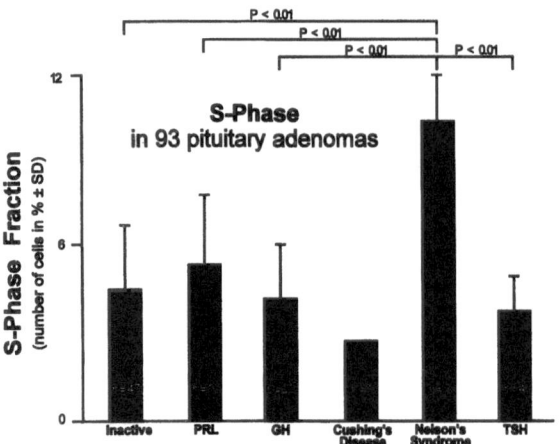

Fig. 4. Flow-cytometry (S-phase fractions) in 93 pituitary adenomas. The adenomas are grouped according to their endocrinological classification

and invasive macroadenomas were compared (Fig. 3). Previously, Knosp *et al.* [10] and Landolt *et al.* [12] showed a similar difference between enclosed and invasive adenomas. Again, pretreatment with dopamine agonists in prolactinomas or somatostatin analogs in acromegalics reduced the proliferative activity. Scanarini *et al.* [15] also demonstrated a reduced percentage of Ki-67 positive cells when growth hormone-secreting tumours were pretreated by octreotide. Furthermore, Knosp *et al.* [10] drew attention to the fact that the percentage of Ki-67 positive cells may vary considerably in different portions obtained from the same tumour. Hsu *et al.* [8] employed PCNA immunostaining to show that recurrent tumours were more rapidly proliferating than those who underwent primary surgery and Shibuya *et al.* [19] confirmed this impression by both α-DNA polymerase and BUdR immunostaining.

DNA-Flow-Cytometry

Alternatively, flow-cytometry is an established method to perform a cell cycle analysis. Additionally, euploid and aneuploid tumours can be differentiated. In slowly growing tumours the rate of cells in the S-phase seems to be particularly important [1]. When the tumour groups were differentiated according to the endocrine classification, tumours with Nelson's syndrome clearly had the highest rates (percentage) of cells in the S-phase (Fig. 4). Nagashima *et al.* [14] confirmed this finding by utilizing a combined in-vitro BrdU infusion and subsequent flow-cytometry approach. Enclosed adenomas were reported to have a significantly lower mean S-phase than invasive tumours [6] and pooled groups of primary adenomas were less rapidly proliferating than recurrent tumours [3].

Conclusions

It appears that the following groups of adenomas harbour an increased proliferative potential: Invasive and recurrent adenomas, large tumours, thyrotropinomas and ACTH-secreting adenomas associated with Nelson's syndrome.

Although these tremendous efforts were able to identify groups which are at higher risk, all the studies have failed to produce reliable results in predicting the prognosis of an individual patient. We have closely followed-up all our patients with unusually high DNA-polymerase and Ki-67 levels but have not found an increased recurrence rate in comparison to those patients with average levels of these parameters. Consequently, decisions as whether to apply radiotherapy or determination of the length of follow-up periods on the basis of studies of tissues resected at surgery should be deferred until further data on this topic become available.

Nevertheless, repeat measurements of proliferation parameters in an individual patient such as the change from Cushing's disease to Nelson's syndrome is able to document the change of biological behaviour.

Acknowledgement

This study was supported by the Johannes and Frieda Marohn Foundation (Buch/90).

References

1. Anniko M, Wersäll J (1988) DNA studies for prediction of prognosis of pituitary adenomas. Adv Biosci 69: 45–52

2. Buchfelder M, Adams EF, Honegger J, Fahlbusch R (1989) DNA polymerase activity in human pituitary adenomas. J Endocrinol Invest 12 [Suppl 2]: 44
3. Chatterjee S, May PL, Forster G, Spiller D, Jeffreys RV (1993) Prediction of recurrence in pituitary tumours: a flow cytometric study using in vivo bromodeoxyuridine. Br J Neurosurg 7: 165–170
4. Deshmukh P, Ramsey L, Garewal HS (1990) Ki-67 labeling index is a more reliable measure of solid tumor proliferative activity than tritiated thymidine labeling. Am J Clin Pathol 94: 192–195
5. Edwards DW, Murthy SR, McGuire WL (1980) Effects of estrogen and antiestrogen on DNA polymerase in human breast cancer. Cancer Res 40: 1722–1726
6. Fitzgibbons PL, Appley AJ, Turner RR, Bishop PC, Parker JW, Breeze RE, Weiss MH, Apuzzo MLJ (1988) Flow cytometric analysis of pituitary tumors. Correlation of nuclear antigen p105 and DNA content with clinical behaviour. Cancer 62: 1556–1560
7. Gerdes J, Lemke H, Baisch H, Wacker HH, Schwab U, Stein H (1984) Cell cycle analysis of a cell proliferation-associated human nuclear antigen defined by the monoclonal antibody Ki-67. J Immunol 133: 1710–1715
8. Hsu DW, Hakim F, Biller BMK, De la Monte S, Zervas NT, Klibanski A, Hedley-Whyte T (1993) Significance of proliferating cell nuclear antigen index in predicting pituitary adenoma recurrence. J Neurosurg 78: 753–761
9. Jefferson G (1952) The invasive adenomas of the anterior pituitary. Thomas, Springfield
10. Knosp E, Kitz K, Perneczky A (1989) Proliferation activity in pituitary adenomas: Measurement by monoclonal antibody Ki-67. Neurosurgery 25: 927–930
11. Kovacs K, Horvath E (1986) Tumors of the pituitary gland. Armed Forces Institute of Pathology, Washington
12. Landolt AM, Shibata T, Kleihues P (1987) Growth rate of human pituitary adenomas. J Neurosurg 67: 803–806
13. Melmed S, Fagin J, Leung M (1986) Bromocriptine inhibits incorporation of [^{3}H] thymidine into rat pituitary tumor cells. Brain Res 369: 83–90
14. Nagashima T, Murovic JA, Hoshino T, Wilson CB, DeArmond SJ (1986) The proliferative potential of human pituitary adenomas in situ. J Neurosurg 64: 588–593
15. Scanarini M (1993) Octeotide therapy of GH-secreting adenomas: histologic and cytological aspects. J Endocrinol Invest 16: 453–454
16. Scheithauer BW, Kovacs KT, Laws ER, Randall RV (1986) Pathology of invasive pituitary tumors with special reference to functional classification. J Neurosurg 65: 733–744
17. Selman WR, Laws ER, Scheithauer BW, Carpenter SM (1986) The occurrence of dural invasion in pituitary adenomas. J Neurosurg 64: 402–407
18. Shaffi OM, Wrightson P (1975) Dural invasion by pituitary adenomas. N Z Med J 81: 386–390
19. Shibuya M, Saito F, Miwa T, Davis RL, Wilson CB, Hoshino T (1992) Histochemical study of pituitary adenomas with Ki-67 and anti-DNA polymerase α monoclonal antibodies, bromodeoxyuridine labeling, and nuclear organizer region counts. Acta Neuropathol 84: 178–183
20. Wiklund JA, Gorski J (1982) Genetic differences in estrogen-induced deoxyribonucleic acid synthesis in the rat pituitary: Correlations with pituitary tumours susceptibility. Endocrinology 111: 1140–1149

Correspondence: Michael Buchfelder, M.D., Neurochirurgische Klinik mit Poliklinik, Universität Erlangen-Nürnberg, Schwabachanlage 6, D-91054 Erlangen, Federal Republic of Germany.

Acta Neurochir (1996) [Suppl] 65: 22–26
© Springer-Verlag 1996

Protein Kinase C and Growth Regulation of Pituitary Adenomas

W.T. Couldwell[1], **R.E. Law**[1,4], **D.R. Hinton**[1,2], **R. Gopalakrishna**[3], **V.W. Yong**[5], and **M.H. Weiss**[1]

Departments of [1]Neurological Surgery, [2]Pathology, [3]Cell and Neurobiology, and [4]Medicine, University of Southern California School of Medicine, Los Angeles, CA, U.S.A., and [5]Department of Neurology and Neurosurgery, McGill University, Montreal, Canada

Summary

The present study was undertaken to explore the role of the Protein Kinase C (PKC) signal transduction system in growth regulation of pituitary adenomas. Primary tumor cultures were plated from fresh surgical tumor specimens. The PKC inhibitors Staurosporine and Tamoxifen were added at varying dosages to the cell cultures. Measurements of cell proliferation were performed by [^{3}H]-thymidine uptake and the [3-(4,5-dimethyl-2-thiazolyl)-2,5-diphenyl-2H tetrazolium bromide] (MTT) assay. After a 48 h treatment period, both [^{3}H]-thymidine uptake and absorbance on the MTT assay decreased in a dose-related manner in both the staurosporine and tamoxifen-treated cultures (IC$_{50}$ of 10 nM and 30 μM respectively). Direct measurement of PKC activity using an in vitro assay revealed very high activity (range of 1465–5708 pmol/min/mg protein; within the range previously published for malignant glioma specimens) in 12 frozen specimens of pituitary adenomas (9 nonfunctional adenomas, 1 prolactinoma, 1 gonadotrophin-secreting and 1 corticotroph-secreting adenoma). In contrast, PKC activity measured in normal adenohypophysis was comparatively very low. These data indicate that pituitary adenoma cells display high PKC activity and are sensitive to growth inhibition by PKC inhibitors. These data suggest a role for the PKC system in regulating pituitary tumor growth, which may have implications for future therapy of these tumors.

Keywords: Protein kinase C; growth regulation; pituitary adenomas.

Abbreviations

FCS – fetal calf serum
PMA – phorbol-12-myristate-13-acetate (phorbol ester; PKC activator)
PKC – protein kinase C
SP   – staurosporine (PKC inhibitor)
TAM – tamoxifen (PKC inhibitor)

Introduction

Phospholipid-dependent protein kinase (PKC) has been demonstrated to play an important role in the transmembrane signaling system in a variety of cell responses to exogenous agonists [19,20]. PKC represents a family of at least 12 known isozymes [17]. Within the pituitary gland, the turnover of phosphoinositides is regulated by neuropeptides and neurotransmitters, which subsequently may generate diacylglycerol (DAG), an endogenous activator of PKC [21]. In this regard, agonist-induced redistribution of PKC activity from cytosol to membranes has been reported in both normal pituitary and adenoma cell lines [7,12,15]. For example, treatment of prolactinoma cells with TRH results in a translocation of PKC activity from the cytosol to membrane, which is followed by an increase in prolactin release [15]. Such work suggests a role of PKC in regulation of hormone release. In support of this concept is the demonstration that phorbol-12-myristate-13-acetate (PMA or TPA, an activator of PKC), stimulates GH secretion from GH-secreting adenoma cells [14,22] and both DAG and phorbol 12, 13-dibutyrate (PDB, another activator of PKC) induce other anterior pituitary hormone secretion in vitro [8,17,18].

Previous work from this laboratory has demonstrated an important role of the PKC signal transduction system in regulating human malignant glioma growth [3–5]. These observations have led to clinical trials utilizing PKC inhibitors as adjuncts in the therapy of patients harboring malignant gliomas. This experience, together with the published data demonstrating the importance of the PKC system in regulating pituitary hormonal function, has prompted the present study in which we explore the PKC system as a growth regulator of pituitary adenoma cells in vitro.

Materials and Methods

Pituitary Adenoma Cell Cultures

For primary human cultured pituitary cells, fresh tumor samples were obtained from a variety of patients with non-functional and functional pituitary tumors. The tumors were removed and plated immediately in 25 cm² tissue culture flasks in DMEM supplemented with 10% fetal calf serum (FCS), penicillin-streptomycin (100 units/ml; 10μg/ml), and 10 mM Hepes buffered to a pH of 7.0 (all medium constituents were purchased from GIBCO, Grand Island, NY). Cells were maintained at 37°C in a humidified 5% CO_2 incubator. Within 3 days of initial plating, the tumors were treated for a 48 h period and then harvested for the proliferation assays ([³H]thymidine uptake and MTT) described below.

Pituitary Adenoma Proliferation Assays

a) [³H]thymidine Uptake Assay

Our method for determining rates of cell proliferation by [³H]thymidine uptake has been previously published [3–5]. In these studies, the PKC inhibitors Staurosporine (SP) [24] (Calbiochem, La Jolla, CA; which was dissolved in ethanol and further diluted in PBS; it was used over a range of concentrations from 1–100 nM) and Tamoxifen (TAM) (Sigma Chemical Co., St. Louis, MO; it was administered over a concentration range of 1–100 μM). Phorbol-12-myristate-13-acetate (PMA, (Sigma), a phorbol ester and activator of PKC [2], was administered over the concentration range of 1–100 nM). These dosages have been previously demonstrated to be within the range known to modulate PKC in vitro [4,5].

b) MTT Assay

As an additional method to quantitate cellular proliferation, the colorimetric MTT [3-(4,5-dimethy1-2-thiazolyl)-2,5-diphenyl-2H tetrazolium bromide] assay was performed following the addition of the PKC inhibitors. Pituitary adenoma cells were seeded at a density of 2×10^3 cells/well in 100 μl feeding medium in flat-bottom 96-well plates (Corning Glass Works, Corning, NY). After 6 h, the test agents were added in predetermined concentrations in quadruplicates to achieve a total volume of 200 μl/well. After incubation for 48 h in the presence of the PKC modulators as above, 20 μl of MTT (Sigma), prepared at a concentration of 5 mg/μl in PBS, was added to each well for 6 h. The plates were spun and supernatant was decanted, and the formazan precipitate was solubilized by addition of 150 μl of 100% DMSO (Sigma) and placed on a plate shaker for 10 min. Absorbance at 570 nm was determined on a Dynatech MR600 spectrophotometric microplate reader.

Tissue Samples for PKC Activity Measurement

For measurement of PKC activity within the histologically verified pituitary adenoma (DRH) or normal anterior pituitary (obtained adjacent to tumor in the removal of microprolactinomas) samples, surgical specimens were snap frozen in liquid nitrogen and stored at –70°C. For assay of PKC activity, tumor or normal anterior pituitary specimens were thawed and approximately 50 mg of tissue was assayed in the manner described below.

Protein Kinase C (PKC) Assay

Pituitary adenoma or normal tissue samples were rinsed twice with ice-cold PBS, followed by a rinse with homogenizing buffer containing 50 mM Tris-HCl, 2 mM dithiothreitol (DTT), 1 mM phenylmethyl-sufonylfluoride (PMSF), and 2 mM ethyleneglycol-bis-(beta-aminoethyl ether) N,N,N',N'-tetraacetic acid (EGTA). The cells were then scraped off the culture plates into 2 ml of the above solution and homogenized (30 strokes) in a glass homogenizer. The homogenate was then centrifuged at 100,000 g for 1 hour. The supernatant was designated the cytosolic fraction. The pellet was resuspended in 2 ml of the above buffer containing 1% Triton X-100, homogenized (30 strokes), and mixed slowly for thirty minutes. This resuspension was then centrifuged at 100,000 g for 1 hour, and the supernatant designated as the particulate fraction. All procedures were performed at 4°C. The enzyme fractions were stored at –70°C prior to assay for PKC.

The method used to assay for PKC activity (ATP transfer into lysine-rich histone) has been previously published [4,5]. When results are expressed as total PKC activity, this represents the sum of the cytosol and particulate activity in the same tissue sample.

Results

Pituitary Adenoma Growth Following Treatment with PKC Modulators

Figure 1 demonstrates the results of a series of experiments in which cultured pituitary adenoma cells were grown in the presence of modulators of PKC for a period of 48 hours. On the [³H]thymidine uptake assay, the PKC inhibitors tamoxifen (Fig. 1 upper) and staurosporine (lower) inhibit uptake in the null cell and prolactin-secreting adenoma cells in a dose related manner. Similarly, the MTT assay shows a decrease in absorbance with increasing concentrations of staurosporine in a primary culture of a pituitary prolactinoma (Fig. 2). In contrast, the administration of the PKC activator PMA increased absorbance on the MTT assay following the treatment period, indicative of an increase in cell numbers, and a proliferative response to treatment with phorbol esters at concentrations which elicit tumor promotion and induce mitogenesis.

Pituitary Adenomas Express High PKC Activity Compared with Normal Adenohypophysis

As shown in Table 1, the PKC activity measured (pmol ATP transferred per min per mg protein) is much higher in pituitary tumor tissue when compared to normal adenohypophysis. Though there was variability among the tumor samples, the activity measured within the tumors was consistently at an order of

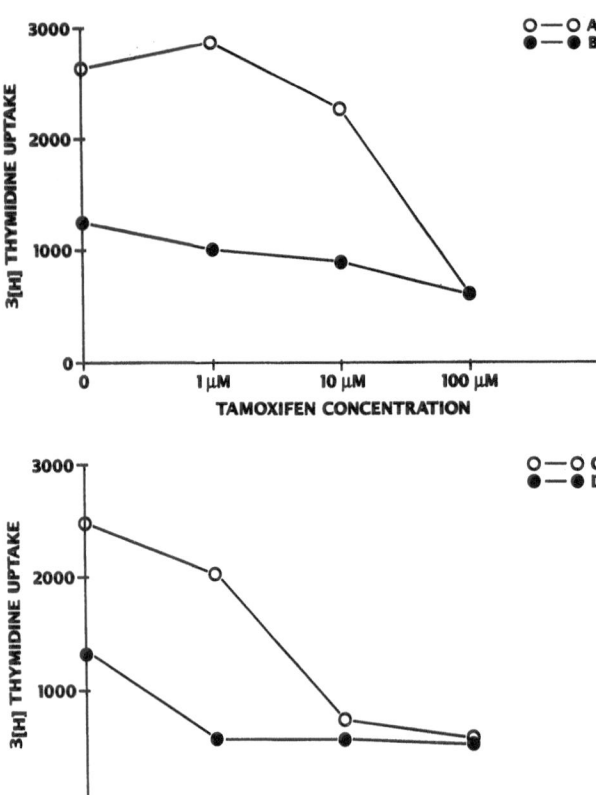

Fig. 1. Pituitary adenoma cell growth inhibition by PKC inhibitors. Administration of the PKC inhibitor tamoxifen (upper) decreases [³H]thymidine uptake in the adenoma cells in a dose related manner. Similarly, staurosporine inhibits thymidine uptake in the adenoma cell within the nanomolar concentration range (lower). Values represent mean of triplicates, with SEM less than 5% of the mean in all cases. Primary tumor cultures *A* and *C* were derived from a prolactin-secreting tumor, and *B* and *D* were derived from a surgical sample of a nonfunctional adenoma

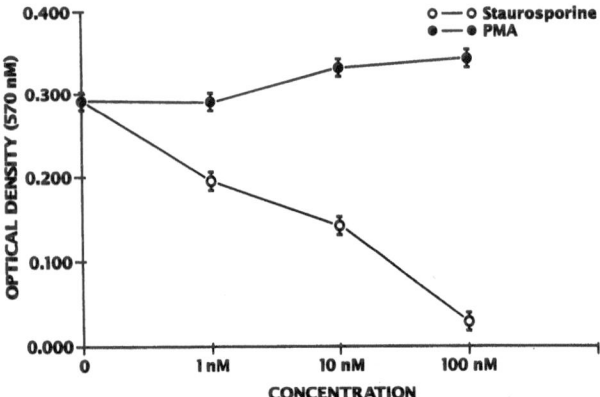

Fig. 2. MTT assay. Diametric effects of the administration of a PKC activator and inhibitor on a cultured PRL-secreting tumor. The administration of the phorbol ester and PKC activator PMA increased absorbance on the MTT assay in the same prolactin-secreting adenoma cell culture that was growth-inhibited by staurosporine. Values are mean of triplicates wells ± SEM

Table 1. *Comparative PKC Enzyme Activity Levels in Pituitary Adenoma Samples and Adenohypophysis*

Cell type	Cell fraction	PKC activity (pmol ATP transferred/min/ mg protein)
1. Nonfunctional	cytosol	1860
	particulate	566
	total	2426
2. Nonfunctional	cytosol	2769
	particulate	890
	total	3659
3. Nonfunctional	cytosol	1096
	particulate	544
	total	1640
4. Nonfunctional	cytosol	1507
	particulate	705
	total	2212
5. Nonfunctional	cytosol	1311
	particulate	418
	total	1729
6. Nonfunctional	cytosol	3280
	particulate	803
	total	4083
7. Nonfunctional	cytosol	1659
	particulate	411
	total	2070
8. Nonfunctional	cytosol	851
	particulate	588
	total	1439
9. Nonfunctional	cytosol	1169
	particulate	488
	total	1657
10. PRL-secreting	cytosol	776
	particulate	689
	total	1465
11. Gonadotrophin-secreting	cytosol	2067
	particulate	170
	total	2237
12. ACTH-secreting	cytosol	4306
	particulate	1402
	total	1640
Adenohypophyseal tissue:		
1.	cytosol	22
	particulate	9
	total	31
2.	cytosol	33
	particulate	0
	total	33

Comparative PKC enzyme activity levels in pituitary adenoma tissue and normal adenohypophysis. PKC functional enzyme levels following isolation of cytosol and particulate fractions of human pituitary adenomas and control non-neoplastic anterior pituitary tissue. Total activity represents the sum of the cytosolic and particulate activity. Equivalent amounts of protein (1 µg) were assayed in all samples. There was markedly elevated PKC levels present in all the adenomas, regardless of functional status, in comparison with the non-neoplastic controls.

magnitude greater than that observed in the normal anterior gland.

Discussion

PKC is a phospholipid-dependent enzyme signal transduction system, which has been implicated in peptide hormone secretion from the pituitary gland [9–11,15]. This work has indicated that PKC may be a key enzyme involved in the control of hormonal secretion. In addition to its role in hormonal secretion, the PKC system has also been implicated in growth regulation, and may have a role in hyperplastic and neoplastic development in a variety of other cell types [4–6,13,19,20,23]. Recent work has suggested that altered PKC activity may play a role in the abnormal growth regulation in human pituitary adenomas [1,15,16]. Immunoreactive PKC expression has been reported to be higher in adenomatous pituitary as compared to normal anterior pituitary tissue harvested postmortem. Furthermore, PKC activity has been noted to be significantly higher in invasive than in noninvasive surgical specimens [1]. In this regard, Todo *et al.* [25,26] have reported higher PKC immunoreactivity in those prolactin-secreting tumors which had poor response to preoperative treatment with dopamine agonists (bromocriptine) than those in which a favorable response had been obtained.

Data from the present study directly measuring PKC activity in a variety of pituitary adenomas is consistent with the previous report by Alvero *et al.* [1] where pituitary adenomas were noted to express high PKC activity. In contrast, the normal human anterior pituitary tissue obtained from surgical specimens in the present study was found to express low PKC activity. The acquisition and PKC measurements in the adenoma and normal pituitary tissue were identical, to enable direct comparison of tissue activity; such methodology was designed to avoid comparison of adenoma with harvested postmortem normal tissue and the inherent potential for degradation of PKC activity over time [1]. The additional demonstration in the present report that inhibition of growth in these tumors may occur with administration of PKC inhibitors, and that addition of PMA, an activator of PKC may increase the growth of these tumor cell in tissue culture further suggests that the elevated PKC activity may in part be responsible for growth regulation of these neoplasms in vitro.

The majority of the PKC activity in pituitary adenomas measured in the present study (Ca^{2+}-depen-

dent classical isozymes, or α-, β- and γ-) is likely to represent that of the α-isozyme (Type III) given the observations of others using immunohistochemical techniques for detection of isozymes among a series of surgical pituitary specimens [16,25]. The ability for the phorbol ester PMA to modulate growth indicates that the PKC isozymes responsible for such control likely belong to the classical or novel groups (see below). Interestingly, Todo *et al.* [25] also found the expression of the α-isozyme to be slight in normal anterior pituitary in comparison to the adenomas, suggesting an alteration of expression of this isozyme may be a component of the transformed phenotype in these cells. Our ongoing studies are directed at the detailed characterization of both classical (i.e., Ca^{2+}-dependent isozymes which bind and are activated by PMA), novel (i.e., Ca^{2+}-independent isozymes) and atypical (Ca^{2+} and PMA independent) PKC isozymes in pituitary adenomas.

In conclusion, the data from the present study indicate that high PKC enzyme activity may be characteristic of adenomatous pituitary. Furthermore, such high activity is amenable to modulation with inhibitors of PKC which produce growth inhibition in vitro. Such high activity may represent, as is currently being evaluated in malignant gliomas, a novel chemotherapeutic target for these tumors.

Acknowledgement

This research was supported by funds from the American Cancer Society (IRG-21-33) and the National Institutes of Health (K08 NS01672-01) (WTC).

References

1. Alvaro V, Touraine Ph, Vozari RR, Bai-Grenier F, Birman P, Joubert D (1992) Protein kinase C activity and expression in normal and adenomatous human pituitaries. Int J Cancer 50: 724–730
2. Castagna M, Takai Y, Kaibuchi K, Sano K, Kikkawa U, Nishizuka Y (1982) Direct activation of calcium-activated, phospholipid-dependent protein kinase by tumor-promoting phorbol esters. J Biol Chem 257: 7847–7851
3. Couldwell WT, Antel JP, Apuzzo MLJ, Yong VW (1990) Inhibition of growth of established human glioma lines by modulators of the protein kinase C second messenger system. J Neurosurg 73: 594–600
4. Couldwell WT, Uhm J, Antel JP, Yong VW (1991) Enhanced protein kinase C activity correlates with the growth rate of malignant human gliomas. Neurosurgery 29: 880–887
5. Couldwell WT, Antel JP, Yong VW (1992) Enhanced protein kinase C activity correlates with the growth rate of malignant gliomas: Part II. Effects of glioma mitogens and modulators of PKC. Neurosurgery 31: 717–724

6. Dekker LV, Parker PJ (1994) Protein kinase C– a question of specificity. TIBS 19: 73–77

7. Drust DS, Martin TFJ (1985) Protein kinase C translocates from cytosol to membrane upon hormone activation: Effects of thyrotropin-releasing hormone in GH_3 cells. Biochem Biophys Res Commun 128: 531–537

8. Emoto N, Ohmura E, Isozaki O, Tsuchima T, Shizume K, Demura H (1991) Phorbol ester, not growth hormone releasing factor, consistently stimulates growth hormone release from somatotroph adenomas in culture. Clin Endocrin 34: 377–382

9. Fearon CW, Tashjian AH Jr (1985) Thyrotropin-releasing hormone induces redistribution of protein kinase C in GH_4C_1 rat pituitary cells. J Biol Chem 260: 8366–8371

10. Fearon CW, Tashjian AH Jr (1987) Ionomycin inhibits thyrotropin-releasing hormone-induced translocation of protein kinase C in GH_4C_1 pituitary cells. J Biol Chem 262: 9515–9520

11. Gordeladze JO, Bjoro T, Ostberg BC, Sand O, Torjesen P, Haug E, Gautvik KM (1988) Phorbol esters and thyroliberin have distinct actions regarding stimulation of prolactin secretion and activation of adenylate cyclase in rat anterior pituitary tumor cells (GH_4C_1). Biochem Pharmacol 37: 3133–3138

12. Hirota K, Hirota T, Anguilera G, Catt KJ (1985) Hormone-induced redistribution of calcium-activated phospholipid-dependent protein kinase in pituitary gonadotrophs. J Biol Chem 260: 3243–3246

13. Housey GM, Houhnson MD, Hsiao WLW, O'Brian CA, Murphy JP, Kirschmeier P, Weinstein IB (1988) Overproduction of protein kinase C causes disordered growth control in rat fibroblasts. Cell 52: 343–354

14. Ikuyama S, Nawata H, Kato KI, Natori S, Ibayashi H (1987) Phorbol ester and phospholipase C-induced growth hormone secretion from pituitary somatotroph adenoma cells in culture: effects of somatostatin, bromocriptine, and pertussis toxin. J Clin Endocrin Met 64: 572–577

15. Ishizuka T, Ito Y, Murayama M, Miura K, Nagao S, Nozawa Y (1987) Hormone-induced redistribution of protein kinase C in human pituitary adenomas. Clin Chim Acta 170: 351–354

16. Jin L. Maeda T, Chandler WF, Lloyd RV (1993) Protein kinase C (PKC) activity and PKC messenger RNAs in human pituitary adenomas. Am J Pathol 142: 569–578

17. Judd AM, Koike K, Yasumoto T, MacLeod RM (1986) Protein kinase C activators and calcium-mobilizing agents synergistically increase GH, LH, and TSH secretion from anterior pituitary cells. Neuroendocrinology 42: 197

18. Negro-Vilar A, Lapetina EG (1985) 1,2-Didecanoylglycerol and phorbol 12, 13-dibutyrate enhance anterior pituitary hormone secretion in vitro. Endocrinology 117: 1559

19. Nishizuka Y (1986) Studies and perspectives on protein kinase C. Science 223: 305–312

20. Nishizuka Y (1988) The molecular heterogeneity of protein kinase C and its implications for cellular recognition. Nature 334: 661–665

21. O'Brian CA, Ward NE (1989) Biology of the protein kinase C family. Cancer Metastasis Rev 8: 199–214

22. Ohmura E, Friesen HG (1985) 12-O-Tetradecanoyl phorbol-13-acetate stimulates rat growth hormone (GH) release through different pathways from that of human pancreatic GH-releasing-factors. Endocrinology 116: 728

23. Persons DA, Wilkison WO, Bell RM, Finn OJ (1988) Altered growth regulation and enhanced tumorigenicity of NIH 3T3 fibroblasts transfected with protein kinase C-1 cDNA. Cell 52: 447–458

24. Tamaoki T, Nomoto H, Takahashi I, Kato Y, Morimoto M, Tomita F (1986) Staurosporine, a potent inhibitor of phospholipid/Ca^{++} dependent protein kinase. Biochem Biophys Res Commun 135: 397–402

25. Todo T, Buchfelder M, Thierauf P, Fahlbusch R (1993) Immunohistochemical expression of protein kinase C type III in human pituitary adenomas. Neurosurgery 32: 635–642

26. Todo T, Shitara N, Nakamura H, Takakura K, Ikeda K (1991) Immunohistochemical demonstration of protein kinase C isozymes in human brain tumors. Neurosurgery 29: 399–404

Correspondence: William T. Couldwell, M.D., Ph.D., USC University Hospital, 1510 San Pablo, Los Angeles, CA 90033, U.S.A.

Acta Neurochir (1996) [Suppl] 65: 27–30

MR Imaging of Residual Tumor Tissue After Transsphenoidal Surgery of Hormone-Inactive Pituitary Macroadenomas: A Prospective Study

P. Kremer[1], **M. Forsting**[2], **J. Hamer**[1], and **K. Sartor**[2]

[1]Neurochirurgische Klinik and [2]Neuroradiologische Abteilung, Kopfklinikum der Ruprecht-Karls-Universität Heidelberg, Heidelberg, Federal Republic of Germany

Summary

22 patients were examined by magnetic resonance (MR) imaging before and after transsphenoidal surgery of hormone-inactive pituitary macroadenomas to evaluate for tumor removal. MR imaging was performed without and with gadolinium-DTPA before the operation and 3 months after. In all cases a suprasellar tumor extension was found preoperatively, in 9 cases with an additional parasellar, in 2 cases with an additional retrosellar extension (average diameter 2.5 cm). In 7 cases complete tumor removal was shown by postoperative MR, but in 11 cases residual tumor tissue was found (4 × suprasellar, 5 × parasellar, 2 × retrosellar). In 4 patients postoperative MR could not clearly differentiate residual tumor from scar formation. Although in cases of residual tumor follow-up MR imaging was performed over a period of two years, residual tumor volumes did not appear to change. This study demonstrates that MR imaging is highly sensitive for evaluating residual tumor tissue after transsphenoidal surgery of hormone-inactive macroadenomas.

Keywords: Pituitary adenomas; magnetic resonance imaging; residual tumor; transsphenoidal approach.

Introduction

Evaluation for complete tumor removal of endocrinologically active pituitary adenomas is commonly possible by measuring postoperative hormone levels. The situation is more difficult after surgery of endocrinologically inactive macroadenomas, which usually are operated on by the transsphenoidal approach [7]. In these cases both the intraoperative impression and the clinical improvement are generally accepted as "gold standard" for the degree of tumor removal. High-resolution MR imaging before and after administration of gadolinium-DTPA is accepted as the most sensitive method for detecting and staging pituitary adenomas [2,4,8,12]. Tumor site, size, extension and relationship to surrounding structures present important information for further therapy. While the interpretation of preoperative MR images of pituitary adenomas is relatively straightforward, the interpretation of postoperative MR images appears problematic and thus has been the center of recent studies [1,13]. Intrasellar changes after surgery induced by scar tissue and tissue implants as well as invasion of the medial wall of the cavernous sinus make interpretation difficult. This study intends to show that the evaluation of residual tumor by postoperative MR imaging after transsphenoidal surgery of hormone-inactive macroadenomas is a valuable method to objectify the degree of tumor removal.

Material and Methods

22 patients (12 men, 10 women, age 29–79 years, average 45 years) were examined by postoperative MR imaging (1.0 T Picker unit). All patients underwent MR imaging before and after administration of gadolinium-DTPA immediately before surgery and 3 months after. All patients had histologically and endocrinologically identified hormone-inactive pituitary macroadenomas, and all were operated on by the transsphenoidal approach. Fat with fibrin glue as implant material was used at the end of surgery in every case. Suprasellar extension of the tumors could be verified in all cases, additional parasellar extension in 9 and retrosellar extension in 2 patients. The average diameter of the tumor was 2.5 cm measured in coronal images (range 1.2 to 4.5 cm). In cases of residual tumor or suspected residual tumor a follow-up of biannual MR imaging was performed for 2 years.

All patients underwent ophthalmological and endocrinological examinations before and after surgery. Clinical symptoms were ophthalmological disorders such as hemianopsia or visual loss (in 14 cases). Four other patients had hormonal dysfunctions and another 4 headache.

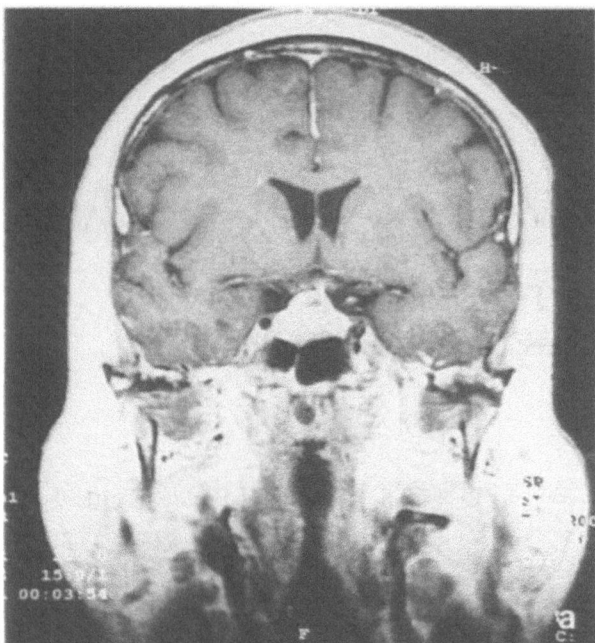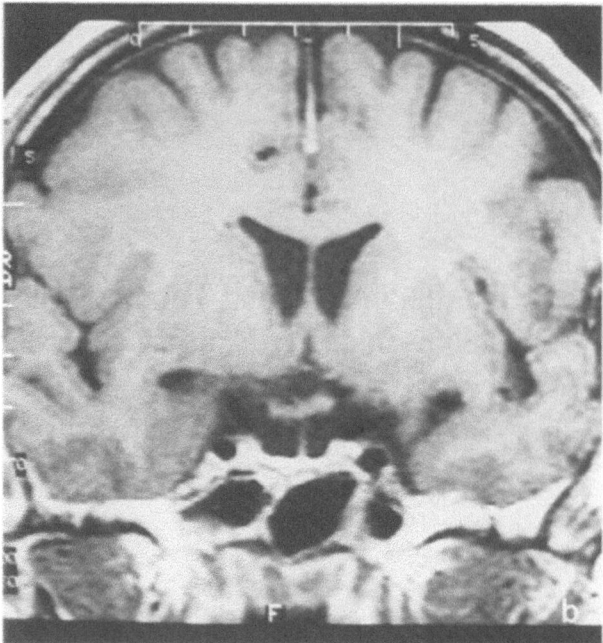

Fig. 1. (a) Contrast enhanced T1-weighted MR image of a pituitary macroadenoma with suprasellar extension and contact to optic chiasma. (b) Postoperative MR image without residual tumor and typical appearance of implanted fat at the sellar floor

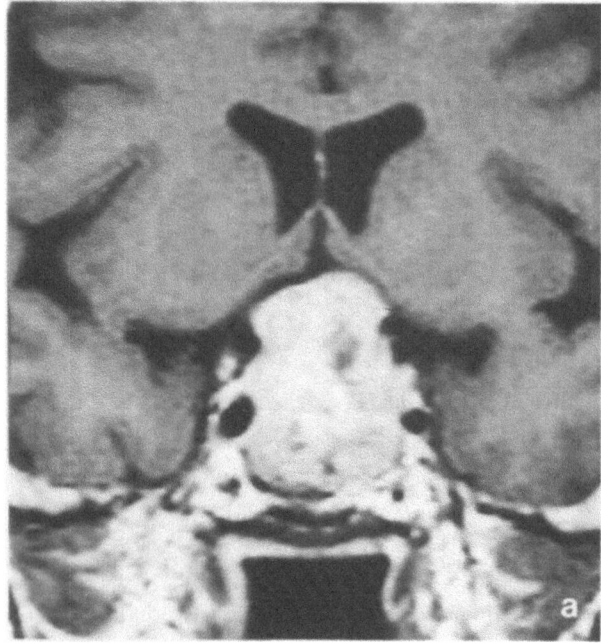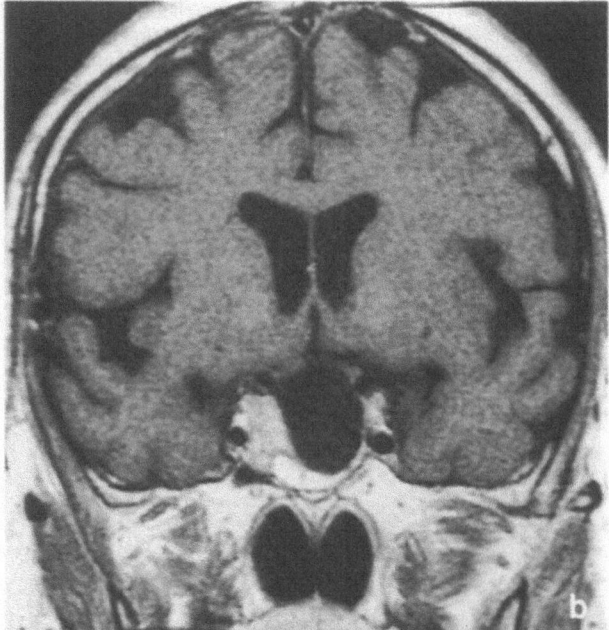

Fig. 2. (a) Preoperative T1-weighted enhanced MR image of a pituitary macroadenoma with parasellar and large suprasellar extension. (b) Postoperative image (same sequence) demonstrating a small residual tumor parasellar on the right side (implant material well delineated at the sellar floor)

Results

Postoperative MR imaging was performed in every patient 3 months after surgery. In 7 patients no residual tumor was seen when preoperative and postoperative MR images were compared (Fig. 1). In 11 patients unequivocal residual tumor (0.6–2.0 cm diameter) was found by MR (Fig. 2) with suprasellar extension in 4 cases, parasellar extension in 5 and retrosellar extension in 2 cases. In 3 of these cases, which had massive

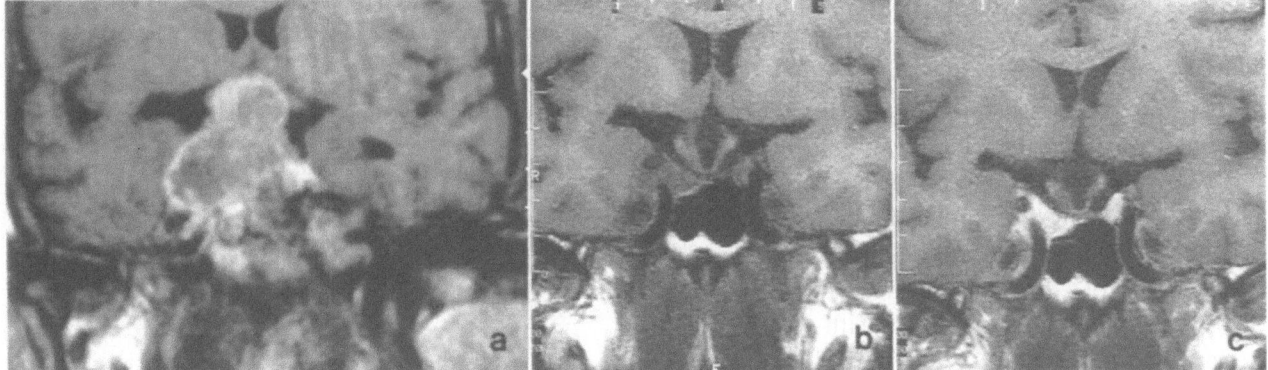

Fig. 3. (a) Contrast enhanced T1-weighted MR-image demonstrating a large supra- and parasellar hormon-inactive pituitary macroadenoma. (b, c) Postoperative images without (b) and with (c) gadolinium-DTPA showing an unspecific enhancing structure along the sellar floor and the near medial wall of the cavernous sinus. Differentiation between scar formation and residual tumor tissue is not possible with certainty

suprasellar extension before surgery, residual tumor was more than 1.0 cm in diameter; all of the other residual tumors were smaller than 1.0 cm. In 4 patients postoperative MR was considered equivocal as residual tumor or scar formation. Enhancing structures at the sellar floor or near the medial wall of the cavernous sinus could not be differentiated with certainty from residual tumor (Fig. 3). Biannual follow-up MR studies over a period of 2 years could not demonstrate any change in the volume of residual tumor or the other enhanced structures.

Evaluation for possible residual tumor was easiest by comparing the pre- and postoperative T1-weighted MR images after paramagnetic enhancement. As to the intraoperative implant material, a distinct decrease could be identified as a small plate with a characteristically high signal in unenhanced T1-weighted images along the sellar floor in 17 of 22 patients.

Complete resolution of ophthalmological symptoms were observed in 9, incomplete resolution in 3 patients. Only 2 patients remained ophthalmologically unchanged. After a period of 2 months of postoperative corticoid substitution this medication could be discontinued in 18 patients.

Discussion

Evaluation for tumor removal in patients with endocrinologically active pituitary adenomas is commonly possible by measuring postoperative hormone levels. Earlier publications from Wilson in 1985 [15] and from Fahlbusch and Buchfelder in 1988 [6] showed that in cases of endocrinologically active macroadenomas persistent hormonal oversecretion could be observed in approximately 30% of growth-hormone-secreting adenomas and in 60% of prolactin-secreting-pituitary macroadenomas. These patients should thus have further medical treatment or radiotherapy to reduce the hormonal oversecretion symptoms. It is generally accepted that in those patients the operative therapy alone does not suffice.

More difficult is the situation in hormone-inactive macroadenomas. Evaluation for tumor removal cannot be measured by endocrinological tests and thus depends largely on the surgeon's intraoperative impression of tumor resection and on the improvement of the clinical symptoms.

In the last years MR imaging has played an important role in the diagnosis of pituitary adenomas and has widely replaced CT scan [9,10]. Pituitary adenomas can be well detected as to size, extension and relationship to surrounding structures by MR imaging with much greater accuracy even for small pituitary adenomas [14]. The interpretation of preoperative MR images of pituitary adenomas has been described in many publications and is well established [2,4,8,12]. More difficult is the interpretation of postoperative images after surgery of pituitary adenomas. Changes in the sella after tumor resection, the use of implanted materials and contrast enhancement of surrounding structures make interpretation more difficult [1,13].

In our prospective study of hormon-inactive macroadenomas we tried to evaluate for presence or absence of residual tumor tissue by postoperative MR done before and after administration of paramagnetic contrast medium. Direct comparison of preoperative and postoperative coronal MR images seems to be best for evaluating the degree of tumor removal. In our series a complete tumor removal was found in 32% of the patients with hormone-inactive macroadenomas. In

50% residual tumor tissue mostly smaller than a diameter of 1.0 cm was unequivocally detected by MR. In 18% of our patients MR did not solve the issue (residual tumor versus scar formation) due to non-specific contrast enhancement at the sellar floor or near the medial wall of the cavernous sinus.

A possible tumor recurrence rate after surgery of pituitary adenomas has been described as being 12% over a period of 4–8 years [3,5]. Even when our follow-up studies with MR in patients with mostly small residual tumor tissues were performed only over 2 years, we never saw a change in the volume of residual tumor. Pituitary adenomas are slow growing tumors [11] and presently it is unclear whether every patient with a small residual tumor tissue will develop a regrowth of the tumor of clinical importance.

We conclude that MR is a highly sensitive imaging method for evaluating residual tumor tissue after transsphenoidal resection of pituitary macroadenoma, especially if the tumor has considerable extrasellar extension. Even if the surgeon's intraoperative impression suggested complete tumor removal, small residual tumor tissue is possible. Postoperative MR should thus complete the assessment by the neurosurgeon. Our observations by postoperative MR show that complete tumor removal after transsphenoidal surgery could be possible only in some patients with pituitary macroadenomas. Further investigations by following-up postoperative MR may be helpful to understand these findings as to what they mean biologically as well as with respect to treatment.

References

1. Bader Ch, Goldmann A, Kunz U, Haeberle HJ, Friedrich JM (1993) Postoperative kernspintomographische Befunde nach Hypophysenadenomentfernung. Fortschr Röntgenstr 159: 476–480
2. Bilaniuk LT, Zimmermann RA, Wehrli FW et al (1984) Magnetic resonance imaging of pituitary lesions using 1.0 to 1.5 T field strength. Radiology 153: 415–418
3. Ciric I, Mikhael M, Stafford T, Lawson L, Garces R (1983) Transsphenoidal microsurgery of pituitary macroadenomas with long-term follow-up results. J Neurosurg 59: 395–401
4. Davis PC, Hoffmann JC, Spencer T, Tindall GT, Braun IF (1987) MR imaging of pituitary adenoma: CT, clinical and surgical correlation. AJNR 8: 107–112
5. Ebersold MJ, Quast LM, Laws ER, Scheithauer B, Randall RV (1986) Long-term results in transsphenoidal removal of nonfunctioning pituitary adenomas. J Neurosurg 64: 713–719
6. Fahlbusch R, Buchfelder M (1988) Transsphenoidal surgery of parasellar pituitary adenomas. Acta Neurochir (Wien) 92: 93–99
7. Hardy J (1969) Transsphenoidal microsurgery of the normal and pathological pituitary. Clin Neurosurg 16: 185–217
8. Kucharczyk W, Davis DO, Kelly WM, Sze G, Norman D, Newton TH (1986) Pituitary adenomas: high-resolution MR imaging at 1.5 T. Radiology 161: 761–765
9. Kulkarni M, Lee KF, McArdle CB, Yeakley JW, Haar FL (1988) 1.5-T MR imaging of pituitary microadenoma: technical considerations and CT correlation. Am J Neuroradiol 9: 5–11
10. Mikhael MA, Ciric I S (1988) MR imaging of pituitary tumors before and after surgical and/or medical treatment. J Comput Assist Tomogr 12: 441–445
11. Nagashima T, Murovic J, Hoshino T, Wilson CB, Dearmond SJ (1986) The proliferative potential of human pituitary tumors in situ. J Neurosurg 64: 588–593
12. Sartor K, Karnaze MG, Winthrop JM, Gado M, Hodges FJ (1987) MR imaging in infra-, para- and retrosellar mass lesions. Neuroradiology 29: 19–29
13. Steiner E, Knosp E, Herold C, Kramer J, Stiglbauer R, Stainszewski K, Imhof H (1992) Pituitary adenomas: findings of postoperative MR imaging. Radiology 185: 521–527
14. Stadnik T, Stevenaert A, Beckers A, Luypaert R, Buisseret T, Osteaux M (1990) Pituitary microadenomas: diagnosis with two- and three-dimensional MR-imaging at 1.5 T before and after injection of gadolinium. Radiology 176: 419–428
15. Wilson C B (1984) A decade of pituitary microsurgery. J Neurosurg 61: 814–833

Correspondence: P. Kremer, M.D., Neurochirurgische Klinik, Kopfklinikum, Ruprecht-Karls-Universität Heidelberg, Im Neuenheimer Feld 400, D-69120 Heidelberg, Federal Republic of Germany.

Acta Neurochir (1996) [Suppl] 65: 31–34

Persistent and Recurrent Hypercortisolism After Transsphenoidal Surgery for Cushing's Disease

U.J. Knappe and **D.K. Lüdecke**

Neurochirurgische Abteilung, Universitätskrankenhaus Eppendorf, Hamburg, Federal Republic of Germany

Summary

After transnasal operations in Cushing's syndrome persisting hypercortisolism either due to negative pituitary exploration or due to subtotal tumor removal, and recurrence of the disease after successful surgery still are challenging. We report on the therapeutic failures among 310 consecutive patients who underwent primary transsphenoidal microsurgery for Cushing's disease. In 287 patients an ACTH-producing pituitary adenoma could be detected (finding rate: 92.6%). In 264 cases remission of hypercortisolism could be attained (remission rate with adenoma 92.0%, for the whole series of primary operations 85.2%). In 23 patients no adenoma could be found despite extensive pituitary exploration (7.4%). Here, we will focus on the management of the 23/287 patients with persistent hypercortisolism after transnasal tumor operation (8.0%) and those 29 cases of the 264 patients with a remission who developed a recurrence of hypercortisolism (11.0%). In recurrent hypercortisolism we recommend transsphenoidal reoperation even when no tumor is visible in MRI. Seventeen of 24 reoperations in recurrent Cushing's disease were successful (70.8%). In persistent hypercortisolism we perform a reoperation during the same hospital stay. Nine of 16 early reoperations led to remission of hypercortisolism (56.3%). If transsphenoidal reoperation fails we indicate radiation therapy of different modalities depending on the extension and location of the tumor remnants. Bilateral adrenalectomy is proposed by us only if all other therapeutic measures failed.

Keywords: Cushing's syndrome; ACTH; pituitary neoplasm; transnasal microsurgery.

Introduction

Transsphenoidal microsurgery is the treatment of choice in endocrinologically proven Cushing's disease. However, in a certain subgroup of patients a longterm cure cannot be achieved. Either negative pituitary exploration, or surgical failure to resolve hypercortisolism in cases where pituitary adenomas could be proven, or recurrences of the disease after primary remission present a major challenge even for very experienced neurosurgeons [4,9,11,12]. With improved preoperative diagnostic tools the incidence of negative pituitary exploration can be decreased [3]. Persistent hypercortisolism is either due to invasive and lateralized microadenomas, where a part of the tumor, e.g. from the cavernous sinus cannot be removed [1,9], or due to macroadenomas with gross invasiveness, where the result of surgery is mostly a subtotal tumor removal rather than a total tumor extirpation. Several endocrinological measures have been proposed to predict recurrences after successful operations [1,7,13,14]. Recurrences have been observed in up to 25% in selected series as long as 13 years after transsphenoidal operation [10]. The aim of this study is to elucidate the management in cases with disappointing outcome after *primary* transsphenoidal surgery for Cushing's disease.

Methods and Material

In a retrospective chart review we evaluated those patients with negative transsphenoidal pituitary exploration (n=23), persisting hypercortisolism after transsphenoidal tumor resection (n=23), and recurrence of hypercortisolism after primarily obtained remission of the disease (n=29), out of our consecutive series of 310 primary transsphenoidal operations in Cushing's syndrome between 1980 and 1991. Surgery was performed by DKL. The mean duration of follow-up was 43.0 months (SD 33.2 months), and was performed by the referring endocrinologists from different centers.

Results

Negative Pituitary Exploration

– In 3 patients with negative pituitary exploration pituitary Cushing's syndrome was assumed, because remission of hypercortisolism was obtained after partial hypophysectomy in one, after external radiation of the sellar region in 2 cases.

– Up to now, in 10 patients an ectopic source of ACTH (n=8) or CRH (n=2) could be proven (7 bronchial carcinoids, one apudoma of the pancreas, one medullary carcinoma of the thyroid gland, one colon carcinoma).
– Ten cases remain unclear. Nine of them underwent bilateral adrenalectomy. Due to improved preoperative diagnostic measures and intraoperative detection of ACTH in peripituitary blood [8] the incidence of negative explorations could be reduced from 12.9% in the first half of the series to 4.1% in the second half. These results will be described elsewhere in further detail (submitted).

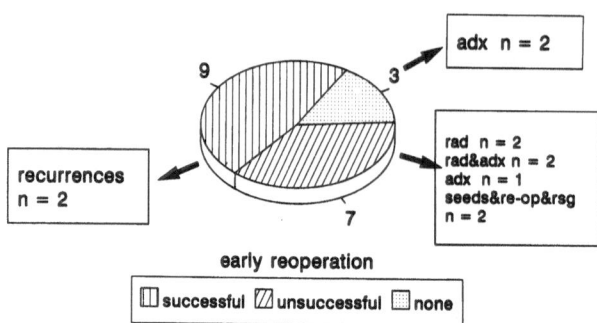

Fig. 1. Management of persistent hypercortisolism after 287 primary transsphenoidal tumor operations, n=19. *adx* adrenalectomy, *rad* conventional radiation, *rsg* radiosurgery

Persistent Hypercortisolism

In persistent hypercortisolemic state we differentiate between two subgroups:

a. Four patients suffered from macroadenomas with gross invasiveness of the surrounding structures. These patients presented with cranial nerve deficits, e.g. diplopia. The endocrinological symptoms were relatively mild. In all 4 patients it was clear, that a surgical cure could not be obtained. The indication for surgery was a reduction of tumor size prior to radiation therapy. After conventional external radiation of the sellar region 2 patients had radiologically no further progress, and 2 patients had a remission of the endocrinological parameters. One of those developed recurrent hypercortisolism and after transnasal reoperation with persistent hypercortisolism eventually underwent bilateral adrenalectomy.
b. The other 19 patients (17 microadenomas, 11 tumors invading either cavernous sinus, 3 invading the neurohypophysis) represent the true surgical failures of this series, because during the primary operation an ACTH-producing pituitary adenoma could be identified and was assumed to be removed. Early postoperative endocrinological evaluation revealed persistence of hypercortisolism and an early reoperation was scheduled in 16 patients within 2 weeks after the first surgery (see Fig. 1). The other 3 patients either refused a reoperation (n=2) or were in a poor clinical state (n=1), two of these patients finally underwent bilateral adrenalectomy.
  – Nine of 16 early reoperations led to a remission of the disease. None of these patients had an insufficiency of the anterior pituitary gland other than the expected transient hypocortisolism.

Therefore the surgical success rate is 95.1% (273/287) in case a tumor can be identified. Two patients sustained a recurrence, one had a second remission after pituitary radiation, the other underwent adrenalectomy, but had to be irradiated later because of the development of a Nelson tumor.
– Seven of the early reoperations were unsuccessful. In 2 patients the disease was under control after external pituitary radiation. Three patients underwent adrenalectomy, after failure of pituitary radiation in 2 cases. The other 2 patients underwent transsphenoidal implantation of ¹²⁵I-seeds, and later radiosurgery. One patient is in remission now. The other patient was in remission, but then developed a recurrence, and was reoperated transnasally in our department. However, this patient died two months after the last operation due to carotid bleeding.

Recurrences

In this series recurrences were observed as long as 10 years after surgery. The incidence of recurrent hypercortisolism after remission is significantly correlated with the decline of the serum cortisol level on the first postoperative morning below 60 μg/l (n=9/160, recurrence rate 5.6%) compared to patients with a decline not below 60 μg/l or normal values (n=20/104, recurrence rate 19.2%. Chi-square, p=0.002). In our clinic hydrocortisone substitution is only started after detection of low serum cortisol level unless arterial hypotension develops earlier. But this is a rare event.

– Twentyfour of the 29 patients with recurrent Cushing's disease have been operated upon a second time, one patient is sheduled for reoperation. The other 4 patients underwent treatment with keto-

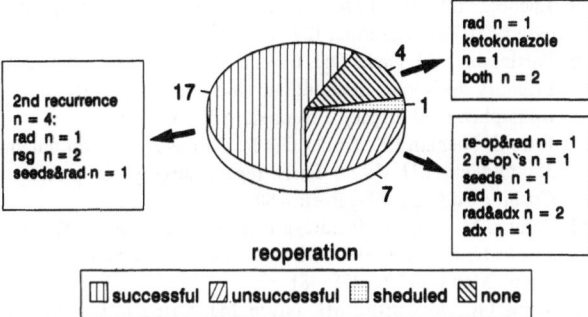

2nd recurrence
n = 4:
rad n = 1
rsg n = 2
seeds&rad·n = 1

rad n = 1
ketokonazole
n = 1
both n = 2

re-op&rad n = 1
2 re-op`s n = 1
seeds n = 1
rad n = 1
rad&adx n = 2
adx n = 1

reoperation

☐ successful ☑ unsuccessful ☐ sheduled ☒ none

Fig. 2. Management of recurrences after 264 remissions, n=29. *adx* adrenalectomy, *rad* conventional radiation, *rsg* radiosurgery

konazole and/or radiation of the sellar region (see Fig. 2).

– Seventeen of 24 reoperations led to remission of hypercortisolism (70.8%). Two patients developed hypopituitarism. Four patients sustained a second recurrence and underwent radiotherapy: conventional external radiation (n=1), after transnasal implantation of ^{125}I-seeds (n=1), or radiosurgery (n=2).

– After 7 of 24 reoperations hypercortisolism persisted. Two patients were reoperated, one of them later on underwent radiating therapy. One remission could be achieved by radiation therapy alone. One patient underwent transnasal implantation of ^{125}I-seeds. Three patients eventually underwent adrenalectomy after 2 radiation failures (see Fig. 2).

Complications

The perioperative morbidity in 310 patients with primary transsphenoidal operations for Cushing's disease is 7.7%. The total mortality in *primary* operations is 0.6% (1 pulmonary embolism, 1 late onset meningitis). The morbidity of transsphenoidal reoperations in this series is 16.7% (2 CSF-fistulas, 2 cases of meningitis, 2 cases of hypopituitarism, 1 persistent diabetes insipidus), there is no mortality.

The above mentioned patient with 2 transnasal adenomectomies, an implantation of ^{125}I-seeds and an external radiation of the sellar region died due to bleeding from one internal carotid artery 2 months after his last operation. Two of 26 patients who subsequently underwent adrenalectomy died due to an Addison crisis.

Discussion

Several endocrinological parameters seem to be predictive for recurrence of Cushing's disease. These include incomplete decline of serum cortisol [7,13] as

suggested by our study, early development of eucortisolism after surgically induced adrenal insufficiency [1] and ACTH hyperresponsiveness to an early postoperative corticotropin releasing hormone (CRH) test [14].

Our results favour a second transsphenoidal operation in recurrent Cushing's disease, and early repeat surgery in surgical failures. As in primary pituitary operations, our goal is selective adenomectomy, if necessary after extensive pituitary exploration, rather than partial hypophysectomy as proposed by others [11]. This explaines the low incidence of hypopituitarism after secondary surgery compared to others [11,12]. Although transsphenoidal surgery is a safe method [6,9], the rate of serious complications is higher in secondary operations, particularly when radiation of the sellar region has been performed.

Radiation is the treatment of choice if pituitary surgery fails to correct hypercortisolism. We refer patients to conventional external radiation either after mass reduction of gross invasive tumors or in cases where the tumor rest can be identified neither during surgery nor in MRI months thereafter. In cases with surgically inaccessible circumscribed tumor rests, e.g. in the cavernous sinus we indicate radiosurgery.

From our experience the most promising issues to improve the results of transsphenoidal microsurgery for Cushing's syndrome are 1) a definitive diagnosis of pituitary dependent hypercortisolism to avoid unnecessary negative pituitary exploration and 2) to obtain better localization even of small microadenomas or remnants of those. The former was improved after the introduction of reliable endocrinological tests and in cases with equivocal results by inferior petrosal sinus sampling (IPSS) [5] in the mideighties as reflected in the decrease of negative pituitary explorations in our series. The latter remains a problem since MRI does not detect minute pituitary adenomas in 38% [5] up to 70% [9] of the cases. Even the invasive IPSS reveals a correct preoperative lateralisation of the tumor only in 76.3% [5]. In vivo labelling of heterotransplanted ACTH-producing pituitary adenomas has been promising [2] but its clinical application is not yet available.

References

1. Buchfelder M, Fahlbusch R (1990) Recurrences in Cushing's disease – prediction and prevention. In: Luedecke DK *et al.* (eds) ACTH, Cushing's syndrome, and other hypercortisolemic states. Raven, New York, pp 281–288

 2. Knappe UJ, Luedecke DK, Puchner MJA, Saeger W, Hermann H-D (1991) In vivo labelling with ^{125}I-CRH of human ACTH-producing pituitary adenomas heterotransplanted to nude mice. Endocr Pathol 2: 200–209
 3. Knappe UJ, Luedecke DK, Saeger W (1993) Negative pituitary exploration in Cushing's syndrome. J Endocrinol Invest 16 [Suppl 1–8]: 129
 4. Kuwayama A (1990) Long term results of pituitary surgery in Cushing's disease. In: Luedecke DK et al (eds) ACTH, Cushing's syndrome, and other hypercortisolemic states. Raven, New York, pp 289–296
 5. Landolt AM, Schubiger O, Maurer R, Girard J (1994) The value of inferior petrosal sinus sampling in diagnosis and treatment of Cushing's disease. Clin Endocrinol 40: 485–492
 6. Laws ER Jr (1990) Complications of surgery for ACTH secreting pituitary tumors. In: Luedecke DK et al (eds) ACTH, Cushing's syndrome, and other hypercortisolemic states. Raven, New York, pp 275–280
 7. Luedecke DK, Niedworok G (1985) Results of microsurgery in Cushing's disease and effect on hypertension. Cardiology 72: 91–94
 8. Luedecke DK (1989) Intraoperative measurement of adrenocorticotropic hormone in peripituitary blood in Cushing's disease. Neurosurgery 24: 201–205
 9. Luedecke DK (1991) Transnasal microsurgery of Cushing's disease 1990. Path Res Pract 187: 608–612
10. Partington MD, Davis DH, Laws E Jr, Scheithauer BW (1994) Pituitary adenomas in childhood and adolescence. Results of transsphenoidal surgery. J Neurosurg 80: 209–216
11. Ram Z, Nieman LK, Cutler GB Jr, Chrousos GP, Doppmann JL, Oldfield EH (1994) Early repeat surgery for persistent Cushing's disease. J Neurosurg 80: 37–45
12. Somma M, Rasio E, Beauregard H, Serri O, Comtois R, Aris-Jilwan N, Hardy J (1993) The treatment of Cushing's disease. Union Med Can 122: 478–481
13. Toms GC, McCarthy MI, Niven MJ, Orteu CH, King TT, Monson JP (1993) Predicting relapse after transsphenoidal surgery for Cushing's disease. J Clin Endocrinol Metab 76: 291–294
14. Vignati F, Berselli ME, Loli P (1994) Early postoperative evaluation with Cushing's disease: usefulness of ovine corticotropin-releasing homone test in the prediction of recurrence of disease. Eur J Endocrinol 130: 235–241

Correspondence: Ulrich J. Knappe, M.D., Neurochirurgische Abteilung, Universitätskrankenhaus Eppendorf, Martinistr. 52, D-20246 Hamburg, Federal Republic of Germany.

Acta Neurochir (1996) [Suppl] 65: 35–36

Surgical Treatment of Pituitary Adenomas in Elderly Patients

J. Pospiech, D. Stolke, and **F.R. Pospiech**

Department of Neurosurgery, University Hospital GHS Essen, Federal Republic of Germany

Summary

Partly due to increased life expectancy, more and more patients over 60 years of age present with neurosurgical problems. In each case you have to decide to operate a patient or not. Describing the management in pituitary adenomas we conclude that also in elderly people operative therapy via the transsphenoidal approach can be done with a low risk in most cases.

Keywords: Complication; pituitary adenoma; transsphenoidal approach.

Introduction

There exist several aspects in the management of pituitary adenomas in elderly patients which are worthwhile to mention. Pituitary adenomas, glioblastomas, and meningiomas are the most common primary brain tumors in older patients [1,2]. The incidence of different histological subtypes as well as clinical symptoms is age-related [3–6]. Moreover in patients over 60 years of age (60y) frequently cardiovascular and pulmonary function is impaired preoperatively, so that the overall operation-related risk is increased [1]. It becomes evident, that the indication for operation therefore must be strongly correlated with good results.

Patients and Clinical Methods

Between 1990 and 1992 we operated on 128 patients with pituitary adenomas. 48 patients were older than 60y (37.5%). General patient characteristics are listed in Table 1. Both subgroups were analysed with regard to clinical presentation, physical examination and concomitant diseases, tumor extension, operative procedure, histology, and perioperative morbidity and mortality. The last follow-up at 8.3 months in the older subgroup and 8.7 months for younger patients included physical examination with a special interest in ophthalmological function.

Results

Patients older than 60y mainly present with visual disturbances, in 62.5% (Table 2). In 65% and 69%, respectively, vision impairment and visual field defects could be diagnosed. Half of the patients showed concomitant diseases (e.g. arterial hypertension, coronary heart disease etc.). In the younger subgroup the most common cause, which led to the diagnosis of a pituitary adenoma, was endocrinological dysfunction (55%). Only 38.8% showed a reduction in vision, in 32.5% visual field defects were observed. In 15 cases (19%) cardiovascular and pulmonary risk factors were known.

73% of the tumors in patients older than 60y were non-secreting adenomas, in 90% a suprasellar extension was noted. Histological evaluation in younger

Table 1. *General Patients Information*

	< 60y	> 60 y
Number	80	48
Sex – m/f	46/34	23/25
Age – mean	43.2	66.2
Recurrence – %	16	31
Follow-up	8.7 mon	8.3 mon

Table 2. *Clinical Symptoms, in Percent*

	< 60 y	> 60 y
Ophthal.	39	62.5
Endocrine	55	28
Neurolog.	25	4
X-ray	20	21

Ophthal. visual field defects and impairment of vision; *endocrine* signs for hormone insufficiency or hypersecretion; *neurolog.* mostly cranial nerve dysfunction or organic psychosyndrome; *X-ray* accidental finding or progression of known residual tumor

Table 3. *Complication Rates, in Percent*

	< 60 y	> 60 y
Minor	37.5	35
Major	7.5	12.5
Perman. def.	2.5	6.25
Mortality	2.5	none

Minor minor morbidity; *major* major morbidity; *perman. def.* with permanent neurological deficits

patients revealed in 46% adenomas without and in 54% tumors with signs of hormone secretion. Only in 70% tumor location was the suprasellar.

The operative approach was transsphenoidal in all patients over 60y, total tumor removal could be achieved in 90%. In 4 of 80 cases in the younger age-group we operated transcranially, mainly because of para- and retrosellar tumor extension. In 25% with infiltration of the cavernous sinus only a subtotal adenoma resection was possible.

Minor perioperative morbidity (e.g. diabetes insipidus, transient CSF leak, meningitis without permanent deficit) was noted in both groups with nearly the same incidence – 35% in older and 37.5% in younger patients. Major morbidity was slightly higher for patients over 60y than under 60y – 12.5% and 7.5%, respectively (Table 3). However, none of the older patients died perioperatively, whereas the mortality rate was 2 of 80 (2.5%) in young patients. In both cases it came to massive postoperative hemorrhage with consecutive hydrocephalus. Overall complications resulted in permanent neurological deficit for 6.25% of patients over 60y and for 2.5% of the other subgroup.

At last follow-up in about 50% of the older and in nearly 60% of the younger patients, the ophthalmological deficits were improved. In all other cases visual disturbances remained stable.

Endocrinological analysis was restricted to non-secreting adenomas, because of the low number in the older age-group. Significant differences with regard

to postoperative incidence of pituitary insufficiency between both groups could not be found.

Discussion and Conclusion

Transsphenoidal surgery seems to be a safe procedure in the management of pituitary adenomas in elderly patients. A significant para- and retrosellar extension could not be noted in our series, so that total tumor removal was possible via the midline approach in most cases. In contrast to younger people [3,6] the tumors are mainly located in the intra- and suprasellar regions and are non-secreting adenomas, which resulted in typical chiasma syndrome. Relief of chiasma compression by adenoma resection led to improvement of ophthalmological function in nearly the same frequency as in younger individuals. Overall complication rate was in accordance with the literature [1] relatively low – 6.25% – in comparison to the younger subgroup – 2.5%.

References

1. Arnold H, Schwachenwald R (1989) Cerebrale Tumoren im Alter. Z Geriatrie 2: 486–491
2. Cervos Navarro J, Ferszt R (eds) Klinische Neuropathologie. Thieme, Stuttgart
3. Davis DH, Laws ER, Ilstrup DM, Speed JK, Caruso M, Shaw EG, Abboud CF, Scheithauer BW, Root LM, Schleck C (1993) Results of surgical treatment for growth-hormone secreting pituitary adenomas. J Neurosurg 79: 70–75
4. Dyer EH, Civit T, Visot A, Delalande O, Derome P (1994) Transsphenoidal surgery for pituitary adenomas in children. Neurosurgery 34: 207–212
5. Partington MD, Davis DH, Laws ER, Scheithauer BW (1994) Pituitary adenomas in childhood and adolescence. Results of transsphenoidal surgery. J Neurosurg 80: 209–216
6. Yamada S, Kovacs K, Horvath E, Aiba T (1991) Morphological study of clinically nonsecreting pituitary adenomas in patients under 40 years of age. J Neurosurg 75: 902–905

Correspondence: Josef Pospiech, M.D., Department of Neurosurgery, University Hospital GHS Essen, Hufelandstr. 55, D-45147 Essen, Federal Republic of Germany.

Acta Neurochir (1996) [Suppl] 65: 37–40

Combined Surgery and Radiotherapy of Invasive Pituitary Adenomas – Problems of Radiogenic Encephalopathy

F. Rauhut and **D. Stolke**

Department of Neurosurgery, University Hospital Essen, Federal Republic of Germany

Summary

The aim of the present study was to find out: can combined modality of surgery and local irradiation in case of invasive pituitary adenomas prevent tumour relapse and develop radiogenic late damage after this therapy. Thirty-three patients suffered from primary hypophysomas and twenty-three patients had pituitary tumour recurrences. In all cases combined therapy was performed. The long term results of 56 patients showed a recurrence rate of 5.4% (3 cases). The median follow-up time was 152 months, ranging from 101 to 197 months. Clincal and neuroradiological signs of radiogenic encephalopathy developed in 16 cases (28%). Because of slow progression, the exact beginning of the neurological symptoms was difficult to determine and varied from 7 to 11 years. The extent of the cerebral and endocrinological disturbances was very different but a marked effect of the optic nerve was not seen. It seems that the usual X-ray therapy with single doses of 1.8 Gy and total doses of 45–50 Gy have a high risk of radiogenic morbidity.

Keywords: Pituitary recurrence; radiotherapy; radiogenic encephalopathy; CT-/MR-findings.

Introduction

The invasive pituitary tumours with infiltration of the neighbouring structures such as the arachnoid membrane, dura and the skull base make it impossible for them to be removed completely. Their high risk of local tumor recurrence is well known. As reported in the literature, there is a 45–51% recurrence rate after surgery. In addition to surgery, many authors performed radiotherapy to prevent tumour relapse postoperatively [4,6,5,10,18]. Such combined treatment reduced the tumour recurrence rate to 2–7% [9,13–16]. Although patients suffered from a progressive pituitary deficiency after radiotherapy, they had a good quality of life in the first few years after irradiation. In our own patients, the late effects of such therapy show a high risk of radiogenic damage of cerebral structures around the tumour area indicative of radiogenic encephalopathy.

Clinical Material and Methods

In the present study, 56 patients with invasive pituitary adenomas were investigated. Thirty-three suffered from a primary tumour and in twenty-three cases a tumour relapse was observed. Hormone-inactive tumours were seen in 34 cases; the remainder were hormone-active adenomas (5 acromegaly, 7 M. Cushing, 10 prolactinomas). In the encephalopathy group, only 3 were prolactinomas. Local irradiation was carried out with two parallel opposed fields using a 5.7 MeV linear accelerator or cobalt-60 system. The total dose varied from 45 to 60 Gy although in most cases 50 Gy were applied in single doses of 1.8–2.5 Gy using five fractions per week. Clinical, endocrinological and neuroradiological controls were performed in all patients.

Results

In only 3 cases (5.4%), tumour relapse was observed after radiotherapy. In 16 patients (28%), the follow-up examination showed clinical and neuroradiological signs of radiogenic encephalopathy. The median follow-up time was 152 months, ranging from 101 to 197 months (Fig. 1). Thirty-seven patients without

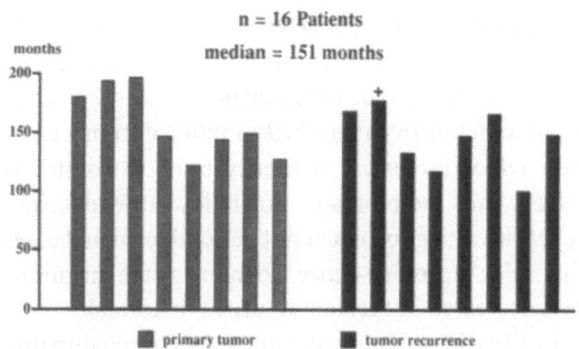

Fig. 1. Follow-up times in patients with encephalopathy, after radiotherapy. n = 16 patients

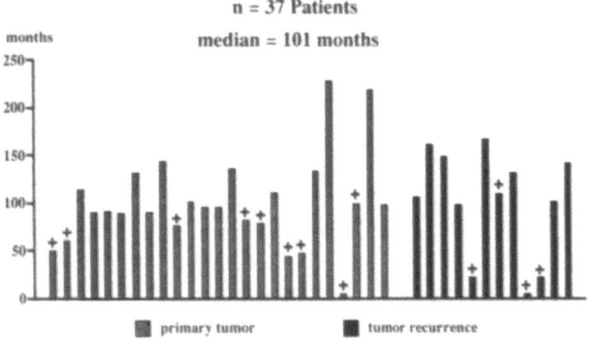

n = 37 Patients
median = 101 months

Fig. 2. Follow-up times in patients without encephalopathy, after radiotherapy. n=37 patients

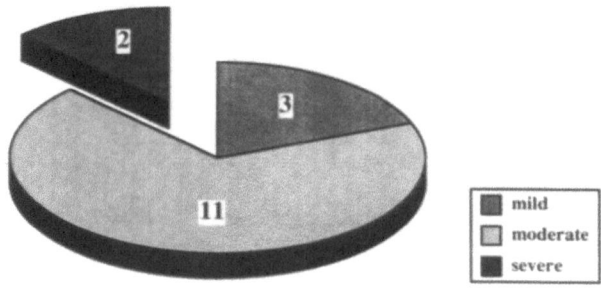

Fig. 3. Grading of the encephalopathy. n=16 patients

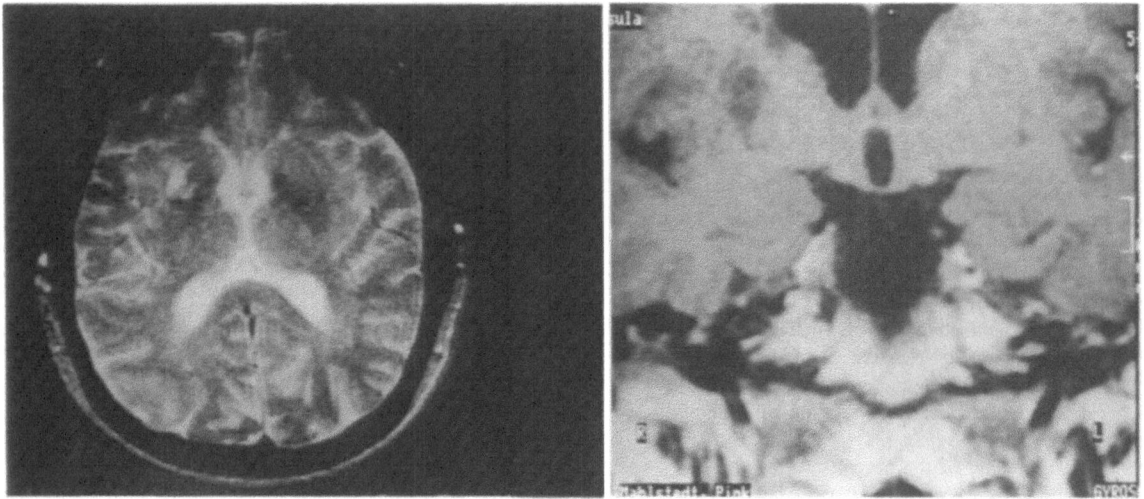

Fig. 4. MRI-scan shows radiation injury in the region of basal ganglia (left T2-weighted, right T1-weighted image)

encephalopathy had a median follow-up time of 101 months varying from 2 to 230 months (Fig. 2). A large percentage (33%; 13 patients) of this group died. Impaired cerebral function and loss of memory were the main symptoms of encephalopathy in all cases. In addition, a few cases developed synergia, cranial nerve dysfunction, aphasia, diencephalic and Parkinson syndrome (Fig. 3). Radiotherapy had no marked effect on the optic nerve. A slight deterioration in visual acuity was observed in only 3 cases (5%). The secretion of the anterior pituitary lobe worsened in 33 cases (30%) after irradiation but 16 patients (29%) suffered from a complete endocrinological deficiency prior to combined surgery and radiotherapy. An follow-up endocrinological investigation could not be conducted in 2 cases only. For hormone-active adenomas, the hormone level decreased but seldom returned to normal.

In 12 cases patients with radiogenic encephalopathy CT- and MRI-scans showed pathological findings such as enlargement of the subarachnoidal space and the

ventricles, white matter lesions, periventricular hypodensity, lacunar lesions in the region of basal ganglia and in the cortical and subcortical areas (Fig. 4). Because of slow progression, the exact beginning of encephalopathy was difficult to determine. In our patients, the symptoms began approximately 7 to 11 years after radiotherapy. The effect of age was investigated in both groups. For the encephalopathic series, the age distribution ranged from 25 to 61 years (median 45 years) at the start of radiotherapy and from 35 to 71 years (median 57 years) at the present time. The age of the patients without encephalopathy varied from 19 to 70 years (median 49 years) at start of therapy and from 29 to 79 years (median 58 years) presently.

Discussion

Although complete removal of pituitary adenomas with the transsphenoidal or transcranial technique is not always possible, prolonged symptom-free intervals can frequently be achieved [5,9,18]. In 1978, Guiot was the first to report on the advantage of postoperative irradiation of invasive pituitary tumours to decrease the frequency of relapse [6]. Studies by Sheline and Tyrell (1983) showed a 5-year determinate recurrence rate of 4% for resection plus radiotherapy [14,16]. In our own patients, we saw no tumour relapse following combined therapy over a period of five years although a recurrence rate of 5% developed later [3,10]. Ebersold *et al.* reported on the effects of radiation therapy in 50 patients. A tumour relapse developed in 18% of these cases. In another group of 42 patients, who did not receive radiation therapy after surgery, showed a recurrence rate amounting to 12%. The authors suggest that irradiation by itself cannot prevent recurrence even when the tumour is significantly debulked [4]. It was discussed whether local X-ray therapy damaged the tumour vascularisation with subsequent inhibition of growth. Morphological investigations on primary pituitary tumours and irradiated pituitary recurrences revealed no changes in the cytological and histological structures of the adenomas [11]. It is thought that the cerebral tissue around the midline process tolerates a total dose of 50 Gy provided the single dose is not too high. Initially the single doses amounted to 2.5 Gy, but for the majority of patients the dose was subsequently reduced to 1.8 Gy. As expected, the function of the anterior pituitary worsened after radiotherapy but it was astonishing that in 21 cases (37%) no decrease in secretion developed. Numerous authors have reported a hormonal disharmony rate of 8–14% but it seems that endocrinological deficiency increases with prolonged follow-up time.

Since a large number of patients had marked visual deficiency prior to start of therapy, there was a danger of additional radiation-induced damage to the chiasma and the optic nerves. The present study shows that the optic system is insensitive to radiation. Visual deficiency increased slightly in only 3 patients but without functional importance. Apart from this, it is questionable whether radiotherapy alone is the reason for the disturbance or whether the residual tumour caused the chiasma compression with deterioration in visual acuity. Various authors have reported significant radiogenic late sequelae of the optic nerves. Harris and Levene 9,1% (single dose 2.5 Gy/total dose 45–50 Gy), Aristazabal *et al.* 3% (single dose 2–2.5 Gy/total dose 50 Gy), Rohloff *et al.* 3% (single dose 2 Gy/total dose 50 Gy) and Atkinson *et al.* 17.2% (single dose 3 Gy/total dose 42–45 Gy). These studies came to the conclusion that single doses

are the crucial factor for damage to the optic apparatus [1,2,7,12].

The corpora mammilaria are located near the X-ray field. In the present study, this anatomical configuration would explain why all our patients with encephalopathy suffered from intensive loss of memory. Disturbance of short-term memory was more frequent than loss of long-term memory. All encephalopathy-group patients suffered from a marked psychological disturbances associated with personality change, emotional intellectual degeneration and lack of drive. The MRI-scan showed lacunar defects in the diencephalon. Such localized cerebral damage is almost certainly the reason for the affective impairment. Defects in the medial cerebral nuclei explain the asynchronism developed in 6 cases. The speech centre is situated peripherally to the x-ray maximum but is directly located in the x-ray field. In 2 cases we have seen infarction in this area corresponding with intermittent aphasia 9 years after radiotherapy. The disturbance to the cerebral nerves was insignificant and it is not certain that diplopia is induced by radiation. The same is doubtful in one case of M. Parkinson. There are two main forms of late radiation injury, which may occur separately or together: focal injury and diffuse white matter lesion. The radiation injury produced an increase in tissue water as a rule. These changes cause decreased signal on T1-weighted images and increased signal on T2-weighted and proton-density images. Enhancement following the administration of paramagnetic contrast material is the result of blood-brain barrier defect. Morphological investigations show mineralizing microangiopathy (arteriolar hyalinazation, fribrinoid necroses, demyelination, coagulation necroses). Late findings after radiotherapy are enlarged sulci and dilated ventricles. Tsuruda *et al.* did not find diffuse white matter injuries in case of irradiated pituitary tumours. Valk and Dillon observed several cases of a moyamoya-like syndrome in young patients treated for craniopharygioma with focal irradiation of the sellar and suprasellar regions [17].

The onset of encephalopathy was difficult to determine because of its slow initiation of psychological symptoms in patients exhibiting only a slight subjective feeling of illness in the late stage of the disease. Therefore, it is necessary to interview family members and persons from within the patients' social field. Nevertheless, the cerebral dysfunctions are difficult to diagnose initially. Individual differences exist but it seems that encephalopathy developed after 8–11 years. However, no such disturbances have been reported in the literature because most studies have not been concerned with such long follow-up periods as in our investigation but only single cases of cerebral necrosis [1,4]. In our patient group, it should be emphasized that out of 14 patients, 12 died without tumour disease, so that the risk of encephalopathy could not be absolutely specified.

Under these circumstances it may be of interest that proton beam therapy leads to a similar radiogenic damage. Kjellberg and Kliman reported good results following such therapy; endocrinological insufficiency developed in only 10% of their cases [8].

The number of cases in the various groups is too small to obtain data with statistical significance. Even so, from our experience radiotherapy of pituitary adenomas needs to be reconsidered in order to avoid irreversible radiogenic damage. Further reduction of the total dose is a problem with respect to tumour recurrence [12]. Young persons in particular should not be irradiated. It remains to be seen whether new radiation modalities, for example, gamma knife or interstitial irradiation can reduce the risk of encephalopathy.

Conclusions

The combined modality of surgery and radiotherapy for pituitary adenomas can reduce the frequency of tumour relapse. Our long term results of 56 patients showed a recurrence rate of 5%. Radiogenic encephalopathy developed in 16 patients (28%) from 8 to 11 years after therapy but no significant damage to the optic nerves was seen. We suggest that single doses of 1.8 Gy and total doses of 45–50 Gy cannot prevent radiogenic damage. The indication for radiotherapy of hypophysomas must be reconsidered and should be restricted to tumours with enlarged invasion.

References

1. Aristizabal S, Caldwell WL, Avila J (1977) The relationship of time-dose fractionation factors to complications in the treatment of pituitary tumours by irradiation. Int J Radiat Oncol Biol Phys 2: 667–673
2. Atkinson AB, Allen IV, Gordon DS, Hadden DR, Maguire CJF, Trimble ER, Lyons AR (1979) Progressive visual failure in acromegaly following external pituitary irradiation. Clin Endocrinol 10: 469–479
3. Bamberg M, Langrock J, Rauhut F, Hoederath A, Sack H (1988) Indikationsstellung und Ergebnisse bei der Radiatiotherapie von Hypohysenadenomen. In: Bamberg M, Sack H (Hrsg) Therapie primärer Hirntumoren. Zuckschwerdt, München, S 139–146
4. Ebersold MJ, Quast LM, Laws ER, Scheithauer B, Randall RV (1986) Long-term results in transsphenoidal removal of nonfunctioning pituitary adenomas. J Neurosurg 64: 713–719
5. Fahlbusch R, Buchfelder M (1988) Transsphenoidal surgery of parasellar pituitary adenomas. Acta Neurochir (Wien) 92: 93–99
6. Guiot G (1978) Consideration on the surgical treatment of pituitary adenomas. In: Fahlbusch R, Werder KV (eds) Treatment of pituitary adenomas. Thieme, Stuttgart, pp 202–218
7. Harris JR, Levene MB (1976) Visual complications following irradiation for pituitary adenomas and craniopharyngeomas. Radiology 120: 167–171
8. Kjellberg RN, Kliman B (1979) A system for therapy of pituitary tumors. In: Kohler PO, Ross GT (eds) Diagnosis and treatment of pituitary tumors. Elsevier, New York, p 234
9. Pötter R, Müller R.-P, Al-Dandashi Ch (1988) Ergebnisse der Strahlentherapie von Hypophysentumoren. In: Bamberg M, Sack H (Hrsg) Therapie primärer Hirntumoren. Zuckschwerdt, München, S 165–169
10. Rauhut F, clar HE, Bamberg M, Benker G, Grote W (1986) Diagnostic criteria in pituitary tumor recurrence – combined modality of surgery and radiotherapy. Acta Neurochir (Wien) 80: 73–78
11. Rauhut F, Clar HE, Nau HE (+), Gerhard L, Bamberg M (1991) A clinical, endocrinological, and morphological study of pituitary tumor recurrence. In: Samii M (ed) Surgery of the sellar region and paranasal sinuses. Springer, Berlin Heidelberg New York Tokyo, pp 267–271
12. Rohloff R, Oettle E, Wendt Th, Oeckler R, Werder K v, Buchfelder R, Fahlbusch R, Lieven H v: Ergebnisse nach postoperativer und Rezidivbestrahlung von Hypophysenadenomen. In: Bamberg M, Sack H (Hrsg) Therapie primärer Hirntumoren. Zuckschwerdt, München, pp 179–182
13. Sheline GE, Wara WM (1982) Radiation therapy of pituitary tumors. In: Youmans JR (ed) Neurological surgery, Vol 5, 2nd Ed. Saunders, Philadelphia, pp 3163–3169
14. Sheline GE, Tyrrell B (1974) Treatment of nonfunctioning chromophobeadenomas of the pituitary. Am J Roentg 120: 553–561
15. Sheline GE, Tyrell B (1984) Pituitary adenomas. In: Phillips TL, Pistenmaa DA (eds) Rad Oncol Ann. Raven, New York, pp 1–35
16. Tindall GT, Barrow DL (1986) Disorders of the pituitary. Mosby, St Louis, pp 410, 428–429
17. Valk PE, Dillon WP (1991) Radiation injury of the brain. AJNR 12: 45–62
18. Wilson CB (1984) A decade of pituitary microsurgery. The Herbert Olivecrona lecture. J Neurosurg 61: 814–833

Correspondence: Friedhelm Rauhut, M.D., Department of Neurosurgery, University Hospital Essen, Hufelandstrasse 55, D-45122 Essen, Federal Republic of Germany.

Acta Neurochir (1996) [Suppl] 65: 41–43
© Springer-Verlag 1996

LINAC-Radiosurgery (LINAC-RS) in Pituitary Adenomas: Preliminary Results

J. Voges[1], **V. Sturm**[1], **U. Deuß**[2], **C. Traud**[1], **H. Treuer**[1], **W. Schlegel**[4], **W. Winkelmann**[2], and **R.P. Müller**[3]

[1]Abteilung für Stereotaxie u. funktionelle Neurochirurgie, [2]Medizinische Klinik II , [3]Klinik für Strahlentherapie, Universität zu Köln, and [4]Deutsches Krebsforschungszentrum Heidelberg, Federal Republic of Germany

Summary

From 8/90 through 4/94, 32 consecutive patients with recurrent pituitary macroadenoma (PA) were treated with LINAC-RS after tumour resection and/or radiotherapy. Single doses ranging from 8–20 Gy (median: 14.5 Gy) were applied in 14 patients with acromegaly, 5 with Cushing's disease, 4 with Nelson tumours, 5 with prolactinomas and in 4 with nonfunctioning PA's. Retrospective analysis of 26 patients with a follow-up of $\geq$ 6 months revealed no significant endocrinologic response in patients with Cushing's disease, Nelson tumour or prolactinoma. In contrast in 12 evaluated patients with acromegaly within 6–36 months after LINAC-RS the median GH- value decreased significantly. In 3 nonfunctioning PA's a tumour volume reduction has been observed. We conclude, that LINAC-RS with moderate single doses might be a safe and beneficial treatment in patients with acromegaly or nonfunctioning PA's resistant to conventional therapy. In Cushing's disease, Nelson tumours or prolactinomas higher doses seem to be required.

Keywords: Linear accelerator; radiosurgery; stereotaxy; pituitary adenoma; acromegaly; nonfunctioning adenoma.

Introduction

Radiosurgery (RS) is an attractive approach for the noninvasive treatment of benign and malignant intracranial tumours. In patients with inoperable PA's recurring after conventional therapy, this technique can be used with tolerable risk. Since the early sixties in PA's high single doses ranging from 40 to 150 Gy have been applied using Co-60 beams of the gamma knife or the Bragg peak of protons [2–4,9]. In the presented clinical study RS has been performed for the treatment of PA's with a modified LINAC-device [1] and moderate single doses.

Methods

After fixation of the patient's head in a modified, CT-compatible Riechert-Mundinger stereotactic frame, an intraoperative CT-ex-amination was performed [8]. The CT-data and images of preoperative MR-examinations were transferred to a computer (VAX station VS 3500, Digital Equip. Corp., USA). By use of special computer programs, the tumour borders were outlined manually at the computer screen [6], and target points, width of the collimator and depth dose distribution were calculated respectively. The treatment plan was controlled – 2- and 3-dimensionally – by displaying any interesting isodose line to the CT- and MR-images [7].

In our system [1,5] beam-convergence is achieved by use of 10 arcs of Linac-table rotation in steps of 18° and rotation of the gantry from 20°–160°. Spherical fields from 5–54 mm in diameter can be irradiated with dose gradients from 10% (large fields) to 20% (small fields) per mm distance from the tumour surface. Treatment parameters are listed in Table 1.

Endocrine and MR-investigations as well as ophthalmological and neurological examinations were performed in all patients prior to LINAC-RS and in 6-monthly intervals after therapy. Pituitary function was assessed according to Table 2.

Patients

In 19 female and 13 male consecutive patients with PA's, LINAC-RS has been performed from 8/90 through 4/94. 30/32 patients had a tumour relapse after neurosurgical resection. 2/32 patients were irradiated conventionally prior to LINAC-RS (dose: 45 Gy and 54 Gy). In 3/32 patients iodine-125 seeds were implanted before radiosurgery using a transsphenoidal approach. The function of anterior pituitary gland was intact in 7/32 patients, 12/32 patients had an insufficiency of one axis and 13/32 patients of more than one axis. Patients characteristics are listed in Table 1.

Results

26 patients with a follow-up time of $\geq$ 6 months (median follow-up: 19 months, range: 6–41 months) have been analysed retrospectively. In 12 evaluated

Table 1. *Patients Characteristics and Treatment Parameters*

Age (yrs)		Grading[a]	No. patients
median	45		
range	16–69	IE	2
		IAE	1
Tumour volume (ml)		ID2E	1
median	4.1	II	3
range	1.1–23.9	IIE	7
		IIAE	5
Minimum dose (Gy)		IIDIE	1
median	14.5	III	1
range	8.0–20.0	IIIE	3
		IIIAE	2
Maximum dose (Gy)		IVDIE	1
median	20.0	AE	1
range	11.4–36.0	E	4
Isocenters (no.)	84		
Isodose[b] (no. pat.)		Functioning adenomas (no.)	
50%	1		
56%	2	GH-secreting	14
60%	13	ACTH-secreting	5
65%	1	Nelson tumour	4
70%	8	Prolactinoma	5
80%	7		

[a] Grading (according to [10]), *A* occupies suprasellar cistern. *D* intracranial (intradural) parasellar extension, designating extension in the anterior (1), middle (2) and posterior (3) fossa. *E* parasellar extension into or beneath cavernous sinus (extradural). *I* sella normal or focally expanded (tumour < 10 mm). *II* sella enlarged, tumour ≥ 10 mm. *III* localized perforation of sellar floor. *IV* diffuse destruction of sellar floor.

[b] Isodose: dose covering the treatment volume in percent of the dose given to the target.

patients with GH-secreting macroadenomas (median tumour volume: 4.0 ccm, range: 1.1–14.6 ccm) the median GH-value was significantly reduced from 17.2 ng/ccm to 5.1 ng/ccm ($p < 0.05$, paired t-test). One patient with Cushing's disease died after microsurgical resection of a tumour relapse (40 months after LINAC-RS). Radiosurgery caused no mortality. Side effects were occasional seizures in 2 patients and a Kleine-Levine-Syndrom in 1 patient due to radiation-induced tissue changes (temporal lobe) and a reduced ACTH-reactibility in one of these patients. Results are listed in detail in Table 3.

Discussion

In this retrospective analysis, patients with acromegaly responded significantly (tumour shrinkage and/or decrease of hormone excess) to LINAC-RS with moderate single doses (<20 Gy). Comparable results have been achieved with much higher doses by Kliman *et al.* [4] as well as by Thorén and coworkers [9]. Using the Bragg peak of proton beams and doses from 120–140 Gy or the Gamma-Knife technique and doses from 70–100 Gy, remission rates of 85% [4] and 75% [9] were reported in the studies mentioned.

Table 2. *Endocrine Investigations*

Pituitary function	Basal values	Combined stimulation tests	
	fT_3, fT_4, IGF-1 estradiol/testosterone	ACTH- (parameter: cortisol) TRH- (parameter: TSH and prolactin) LHRH- (parameter: LH and FSH)	
Functioning adenomas	Acromegaly	Cushing	Prolactinoma
	hGH-profile, hGH after TRH stimulation IGF-1	ACTH cortisol profile	prolactin profile prolactin response to TRH

Table 3. *Results after LINAC-RS of 26 Patients with Pituitary Macroadenoma*

Treatment group	No. eval. patients	Follow-up (mths)	Hormone excess[a] pre → post	Tumour volume (% volume reduction compared to intraop. values)
Acromegaly	12	median: 14 range: 6–40	GH: 17.2 → 5.1 (median)[b] IGF: 793 → 591 (median)	median: 50%
Cushing's disease	4	22–40	no significant response	
Nelson tumours	2	6, 30	no significant response	
	1	30	ACTH: 2352 → 591	62%
Prolactinoma	1		lost	
	1	32	prolact.: 217.0 → 16.4	62%
	1	8	prolact.: 2146.0 → 1368.0	not evaluable
	1	21	no significant response	
Non-functioning	3	19–29	–	34%–54%

[a] ACTH: pg/ccm, GH, IGF-1, prolactin: ng/ccm.

[b] Statistically significant: $p < 0.05$.

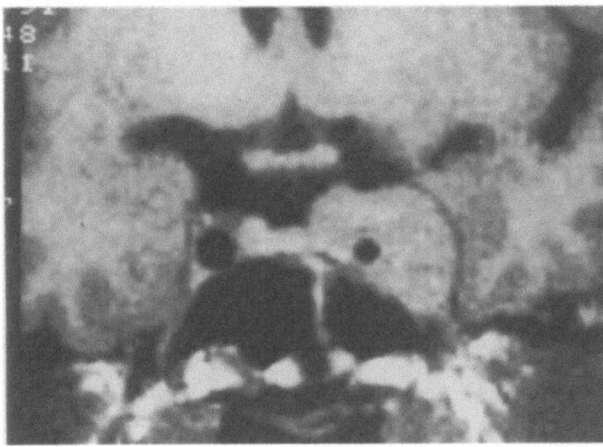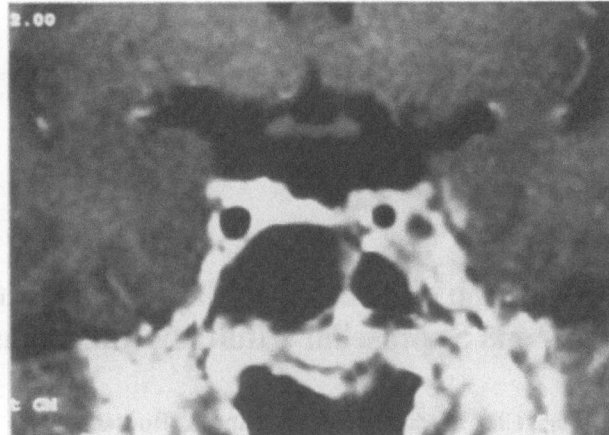

Fig. 1. Patient with a Nelson tumour, grade ID2 E, before (left) and 29 months after LINAC-RS (right) with a single dose to the tumour surface of 15 Gy (80% isodose)

Our results in non-functioning PA's are promising as well. All 3 evaluated patients had a partial response (> 25% tumour volume reduction) after the application of a moderate single dose. The differences in dose-levels, necessary for a significant tumour response, might be due to the advances in treatment-planning and dose-application techniques achieved during the past years.

Except for a few cases, the low doses, which we used had not been sufficient to influence ACTH-secreting adenomas, Nelson tumours or prolactinomas. Two factors can explain these therapeutic failures: 1) Experiences in radiotherapy indicate that the latency period between treatment and response can be long and variable. Thus the follow-up of our patients might have been too short for the detection of a positive effect on the hormone overshoot. 2) ACTH-secreting adenomas, Nelson tumours or prolactinomas might be less radiosensitive than GH-secreting adenomas.

References

1. Hartmann GH, Schlegel W, Sturm V, Bernd K, Pastyr O, Lorenz WJ (1985) Cerebral radiation surgery using moving field irradiation at a linear accelerator facility. Int J Radiation Oncol Biol Phys 11: 1185–1192

2. Kjellberg RM, Kliman B (1979) Proton radiosurgery for functioning pituitary adenoma. In: Tindall GT, Collins WF (eds) Clinical management of pituitary disorders. Raven, New York, pp 315–334

3. Kjellberg RM, Kliman B, Swisher B, Butler W (1984) Proton beam therapy of Cushing's disease and Nelson's syndrom. In: McL Black P (ed) Secretory tumours of pituitary gland. Progr Endocrine Res Ther 1: 295–307

4. Kliman B, Kjellberg RM, Swisher B, Butler W (1984) Proton beam therapy of acromegaly: a 20-year experience. In: Black P (ed) Secretory tumours of the pituitary gland. Progr Endocrine Res Ther 1: 191–21112

5. Pastyr O, Hartmann GH, Schlegel W, Schabbert S, Treuer H, Lorenz WJ, Sturm V (1989) Stereotactically guided convergent beam irradiation with a linear accelerator: localization-technique. Acta Neurochir (Wien) 99: 61–64

6. Schlegel WJ, Scharfenberg H, Sturm V, Penzholz H, Lorenz WJ (1981) Direct visualization of intracranial tumours in stereotactic and angiographic films by computer calculation of longitudinal CT-sections: a new method for stereotactic localization of tumour outlines. Acta Neurochir (Wien) 58: 27–35

7. Schlegel W, Scharfenberg H, Doll J, Hartmann G, Sturm V, Lorenz WJ (1984) Three dimensional dose planning using tomographic data. In: IEEE Comp Society (eds) Proceedings of the Eighth International Conference on the Use of Computers in Radiation Therapy. IEEE Comp Soc Press, Silver Spring, pp 191–196

8. Sturm V, Pastyr O, Schlegel W, Scharfenberg H, Zabel HJ, Netzeband G, Schabbert S, Berberich W (1983) Stereotactic computer tomography with a modified Riechert-Mundinger device as the basis for integrated neuroradiological investigations. Acta Neurochir (Wien) 68: 11–17

9. Thorén M, Rähn T, Guo WY, Werner S (1991) Stereotactic radiosurgery with the cobalt-60 gamma unit in the treatment of growth hormone-producing pituitary tumours. Neurosurgery 29: 663–668

10. Wilson CB (1979) Neurosurgical management of large and invasive pituitary tumors. In: Tindall GT, Collins WF (eds) Clinical management of pituitary disorders. Raven, New York, pp 335–342

Correspondence: Dr. med. J. Voges, Neurochirurgische Universitätsklinik, Abteilung für Stereotaxie und funktionelle Neurochirurgie, Josef-Stelzmann-Str. 9, D-50931 Köln, Federal Republic of Germany.

Acta Neurochir (1996) [Suppl] 65: 44–49
© Springer-Verlag 1996

Comparison of Thermoregulatory Characteristics of Patients with Intra- and Suprasellar Pituitary Adenomas

R. Behr[1], Ch. Dietrich[1], K. Brück[2], and K. Roosen[1]

[1]Neurochirurgische Klinik- und Poliklinik, Julius-Maximilians-Universität Würzburg and [2]Physiologisches Institut, Justus-Liebig-Universität Giessen, Federal Republic of Germany

Summary

Thermoregulatory capabilities under physiologic cold and heat exposure of 37 patients with suprasellar pituitary adenomas (As) and 10 patients with intrasellar adenomas (Ai) were analyzed and compared to each other and to 13 controls (Ctr.) In Ai no shift of the thermoregulatory threshold temperatures was observed. In As the regulation was shifted to a 0.5°C higher mean body temperature in 82% of the patients, indicating a "set-point" elevation. The accuracy of the regulation against thermal loads was maintained, the velocity was reduced. Postoperative examination of As revealed a normalisation of the "set-point". Modifications of the hypothalamic amine systems by the compressive effect of the suprasellar adenomas are discussed to be the most probable cause for the observed thermoregulatory alterations.

Keywords: Temperature regulation; pituitary adenoma; threshold temperatures; noradrenaline; serotonine.

Introduction

The body temperature of patients with pituitary adenomas generally shows normal values. In clinical practice less than 5% of the patients complain of cold intolerance [13]. Lausberg [14] reported no alterations of body core temperature in these patients during external heat and cold load. The skin temperature, however, was reduced. In a recent study [1] an elevation of the tympanic temperature (Tty), a reduction of the mean skin temperature ($\overline{T}$sk) and moreover significant changes of regulatory characteristics were described in patients with suprasellar pituitary adenomas. The aim of this investigation was to confirm the alterations of thermoregulation in a greater population of patients with suprasellar adenomas and to compare them to patients with intrasellar adenomas.

Patients and Methods

37 patients (18 male and 19 female) with suprasellar pituitary adenomas (As) and 10 patients (3 male and 7 female) with intrasellar adenomas (Ai) were examined. 13 healthy subjects served as controls. Anthropometric data are summarized in Table 1.

Endocrinology: In 9 cases of Ai prolactin was increased to 291 ng/ml (SEM 96.04). TSH, T3, T4 and cortisol were in a normal range. In 22 patients of the As group, prolactin was elevated to 1627.6 ng/ml (SEM 666.5). Basal and stimulated values of TSH and cortisol were normal, stimulated TSH was in the lower part of the normal range.

Thermoregulatory methods: Examinations were performed in the climatic chamber of the Department of Physiology, University of Giessen. During external cold and heat load, oxygen consumption (metabolic rate MR), electrical muscle activity (EMA), local sweat rate (SR), local skin blood flow ($\Delta\lambda$), body core (tympanic) temperature (Tty) and mean skin temperature (recorded from 4 sites, $\overline{T}$sk) were measured continuously. Mean body temperature ($\overline{T}$b) was calculated from Tty and $\overline{T}$sk. $\overline{T}$b was plotted against the effectors of temperature regulation MR, EMA, SR and $\Delta\lambda$. The threshold temperatures for the activation of each thermoregulatory effector were

Table 1. *Anthropometric Data of Patients and Controls*

		CTR	Ai	As-i	As-n
Age (years)	mean	32.4	40.6	46.1	45.1
	SEM	2.6	4.9	3.2	4.4
Height (cm)	mean	177.2	172.0	167.6[a]	168.6[b]
	SEM	2.3	3.4	1.5	4.2
Weight (kg)	mean	70.9	94.4[a]	76.3	78.4
	SEM	2.7	5.8	3.5	5.8
Isol (m²/kg 10⁻⁴)	mean	269.7	223.9[a]	252.6[b]	246.3[b]
	SEM	5.1	8.9	5.2	8.6

Anthropometric data of controls (Ctr) and patients with intrasellar adenomas (*Ai*). Suprasellar adenomas with elevated thresholds (*As-i*) and normal threshold (*As-n*). Isol is a measure of the body insulation, the quotient of body surface and body mass. A low value indicates a high insulation. [a] $p < 0.01$, [b] $p < 0.05$ compared to controls.

determined at the breakpoint of the several plots. The slopes of the curves, minimal and maximal values were determined (for detail see [1–3,5]. Statistical analysis was performed by ANOVA, t-test and Wilcoxon).

Ethics: The protocol of this study was submitted to the local ethics committee and has been fully approved. The informed consent of the patients and controls was obtained.

Results

1. Intrasellar Pituitary Adenomas

In thermoneutral ambient temperature (Ta) of 30°C there was no significant difference of body core or skin temperatures. Metabolic rate related to body surface was significantly reduced compared to controls (see Table 2 and Fig. 1).

During cold load, shivering and augmentation of metabolism in order to increase heat production were activated at comparable $\overline{T}b$ – there was no significant shift of the threshold temperatures for heat production ($\overline{T}b$-MR, $\overline{T}b$-EMA). The slope of electrical muscle activity was significantly reduced, indicating a slower reaction of this effector to a physiological cold load.

Heat stress induced sweating at a $\overline{T}b$ comparable to controls. There was also no significant shift of the vasodilation threshold ($\overline{T}b$-λ). However, the slope of the sweat rate curve was significantly reduced by 0.73 W/mK.

Figure 1 illustrates the threshold temperatures and slopes of the thermoregulatory effectors plotted against $\overline{T}b$. The inter-threshold zone is not altered compared to controls (shaded area on x-axis).

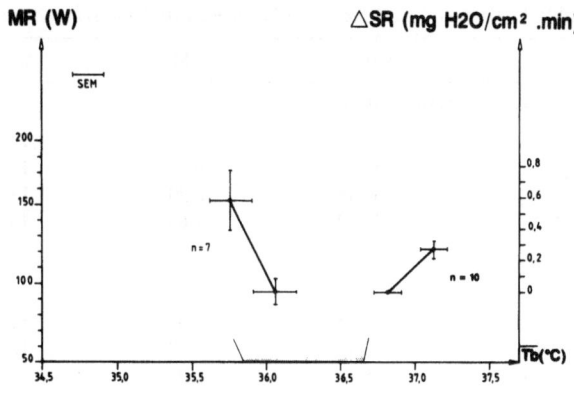

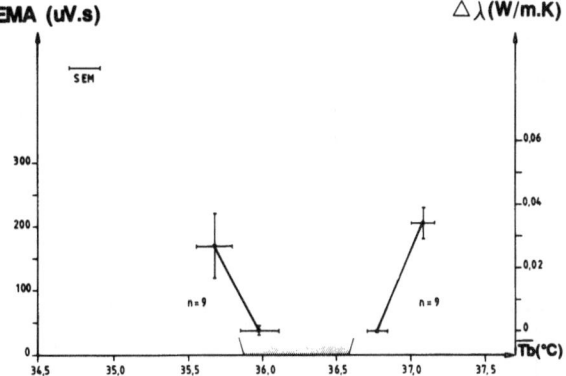

Fig. 1. Thermoregulatory thresholds and curves of patients with intrasellar adenomas (Ai). The base of each curve depicts the threshold temperature of the thermoregulatory effector. The top of the curve is the calculated value (regression line) 0.3°C below or above the threshold respectively. The shaded area on the x-axis is the inter threshold zone of the controls. The small lines left and right the shaded area represents the slope of the control curve. Ai had no significant threshold shifts compared to controls. The slope of the sweating rate (Δ-SR) was significantly reduced, as well as the maximal values during heat load. *MR* metabolic rate; *EMA* electrical muscle acitivity; *SR* sweat rate; Δλ skin heat conductivity = skin blood flow

Table 2. *Thermophysiological Data of Patients with Intrasellar Adenomas (Ai) and Controls (Ctr)*

	Ctr	SEM	n	Ai	SEM	n	p
Neutral ambient temperature							
Ta	30.14	0.44	13	29.97	0.14	10	
Tty	36.78	0.06	13	36.84	0.10	10	
$\overline{T}$sk	34.45	0.20	13	34.13	0.16	10	
MR	48.09	1.63	13	43.12	1.61	10	+
Cold load							
Ta	9.82	1.00	13	13.31	2.21	9	
$\overline{T}$b-MR	35.85	0.07	12	36.06	0.15	7	
Δ-MR	63.94	12.70	12	57.04	1.81	7	
$\overline{T}$b-EMA	35.88	0.07	13	35.98	0.18	9	
Δ-EMA	293.55	48.90	13	131.17	50.40	9	*
Heat load							
Ta	53.57	1.00	13	51.51	0.91	10	
$\overline{T}$b-SR	36.66	0.08	13	36.82	0.09	10	
Δ-SR	0.99	0.28	13	0.26	0.06	10	*
$\overline{T}$b-λ	36.59	0.06	13	36.78	0.07	9	
Δ-λ	0.05	0.01	13	0.03	0.00	9	

$\overline{T}b$-*xx* threshold temperatures. *Δ-xx* slope of the curves. *MR* metabolic rate. *Ta* ambient temperature. + p < 0.05. * p < 0.01 (t-test/Wilcoxon).

Table 3. *Thermophysiologic Data of Patients with Suprasellar Adenomas (As) and Controls (Ctr)*

	Ctr	SEM	n	As	SEM	n	p
Neutral ambient temperature							
Ta	30.14	0.44	13	29.60	0.19	37	
Tty	36.78	0.06	13	37.06	0.04	37	*
$\overline{T}$sk	34.45	0.20	13	33.88	0.12	37	+
MR	48.09	1.63	13	41.86	1.21	35	*
Cold load							
Ta	9.82	1.00	13	15.10	0.81	36	*
$\overline{T}$b-MR	35.85	0.07	12	36.24	0.05	32	*
Δ-MR	63.94	12.70	12	43.71	5.97	32	
$\overline{T}$b-EMA	35.88	0.07	13	36.28	0.05	36	*
Δ-EMA	293.55	48.90	13	200.02	35.40	36	+
Heat load							
$\overline{T}$a	53.57	1.00	13	52.85	0.63	36	
$\overline{T}$b-SR	36.66	0.08	13	36.99	0.04	35	*
Δ-SR	0.99	0.28	13	0.41	0.04	35	*
$\overline{T}$b-λ	36.59	0.06	13	36.89	0.05	29	*
Δ-λ	0.05	0.01	13	0.04	0.01	29	

Legend see Table 2.

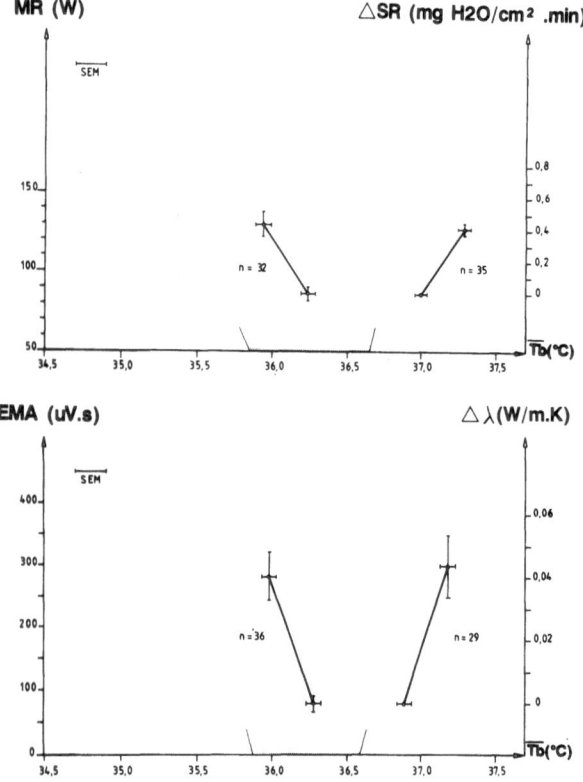

Fig. 2. Thermoregulatory curves of patients with suprasellar adenomas (As). All thresholds are significantly shifted to a higher mean body temperature. The "set-point" is elevated. The inter-threshold zone is not altered, but the slopes are reduced

2. Suprasellar Pituitary Adenomas

Under thermoneutral conditions Tty was significantly increased in As by 0.28°C compared to controls (see Table 3). Mean skin temperature and metabolic rate were significantly lowered by 0.52°C and 6.23 W respectively.

During cold load the patients started heat production at a much higher ambient temperature than controls. The threshold temperatures for heat production were significantly increased by 0.4°C. The slopes of both curves, Δ-MR and Δ-EMA, were reduced (see Fig. 2).

In warm environment sweating was elicited in As at a 0.33°C higher mean body temperature than in Ctr. The slope of the sweating curve was also significantly reduced. Heat induced vasodilation occurred at a $\overline{T}$b increased by 0.3°C compared to Ctr.

Figure 2 depicts a significant shift of all threshold temperatures to a higher level of $\overline{T}$b. The inter-threshold zone is not changed.

Table 3 and Fig. 2 represent the whole group of As. However, in 7 patients (18%) no increase of threshold temperatures was measured. Excluding these patients from analysis, the thresholds in the remaining patients would have been approx. 0.1°C higher: $\overline{T}$b-MR=36.35, $\overline{T}$b-EMA=36.39, $\overline{T}$b-SR=37.03 and $\overline{T}$b-λ=36.96°C.

Postoperative examinations were performed in 10 patients, 3 with normal preoperative thresholds and 7 with elevated preoperative thresholds. In the normal group no changes were seen postoperatively. In 7 patients with preoperatively increased thresholds a

Three patients were examined pre- and postoperatively. Body temperatures and thermoregulatory threshold temperatures were not altered significantly after the operation.

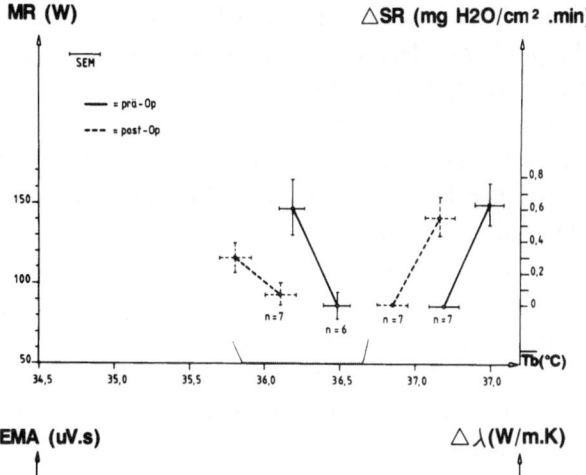

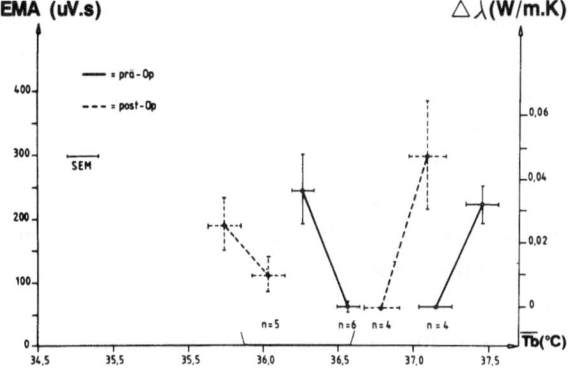

Fig. 3. Thermoregulatory curves pre- and postoperatively of As-i patients. After decompression of the hypothalamus a significant reduction and normalisation of the threshold temperatures was measured. The slopes were not significantly changed

normalisation was observed (Fig. 3). The anthropometric data of the patients was not altered from pre- to postoperative. All threshold temperatures were significantly reduced. In thermoneutrality tympanic temperature was significantly decreased and mean skin temperature increased compared to preoperative values ($p < 0.05$). Resting metabolic rate was significantly increased from 38.02 to 45.36 W/m^2. The slope of the EMA curve was significantly reduced, the others were slightly reduced or unchanged to preoperative measurements.

Discussion

In a recent study [1] modifications of thermoregulatory characteristics in 12 patients with suprasellar pituitary adenomas were reported. In a greater population of 37 patients these early results were confirmed. 82% of the patients with suprasellar adenomas, which compressed hypothalamic tissue and elevated the floor of the third ventricle, had increased threshold temperatures for heat production and heat loss. The inter-threshold zone, describing the accuracy of the central

regulatory system, was not widened. However, the slopes of the curves were reduced. This shows that the velocity of the reactions against thermal loads was diminished. The body core temperature of the As with increased thresholds (As-i) was significantly elevated, despite a reduced resting metabolic rate. As discussed earlier [1], it is assumed, that by reduction of heat loss the reduced mean skin temperature compensates the slightly lowered metabolic rate. Table 1 shows that As had a better thermal insulation, which supports the heat conserving mechanisms.

In 7 patients (18%) no threshold shifts (As-n) were detectable although they had large suprasellar tumors. The anthropometric data of these patients were not different from those of the As-i. The As-n had normal tympanic and mean skin temperatures and a slightly, but not significantly decreased MR in thermoneutral ambient temperature. This gives further support to the above mentioned mechanism of heat storage in As-i. One important thermoregulatory parameter, however, was altered in As-n: the slope of the sweating rate Δ-SR was significantly reduced. The slopes of the other curves were only slightly lowered. The reason for normal thresholds in As was thought to be a lower prolactin level in As-n than in As-i: 161 ng/ml compared to 1627 ng/ml in As-i. However, there was no significant correlation between the thermogenetic thresholds and prolactin values. Furthermore, in patients with fronto-basal meningiomas the same pattern of thermoregulatory modifications was observed, e.g. raised threshold – and body temperatures and reduced oxygen consumption [2]. These meningioma patients had no prolactin secretion.

Patients with intrasellar adenomas (Ai) had a small and insignificant shift of all thermoregulatory threshold temperatures. Body core- and skin-temperatures were not changed. Resting metabolic rate was significantly reduced at Ta 30°C averaging in the range of the MR of As-n. The MR of As-i was still lower. Of special interest is that the slope of the sweat rate curve was extremely low and the maximal sweat secretion was reduced. The reduction of these parameters was similar to patients with diabetes insipidus [3] although none of Ai patients developed a diabetes insipidus clinically. Assuming that intrasellar adenomas cause an increase of the intrasellar tissue pressure [15], it is reasonable that the transport of ADH to the posterior lobe of the pituitary is hindered. This may result in a relative reduction of peripheral ADH causing a lowering of the sweat secretion rate. When the tumor breaks through the diaphragma sellae, the intrasellar pressure may be released and the ADH secretion is facilitated. In As the sweat secretion rate was improved compared to Ai. However, there are no data which would support this theory, except the diabetes insipidus patients, who had a similar pattern of regulation.

The main difference between Ai and As is the altered "set point" of temperature regulation in As-i. In As-i body core temperatures and thresholds are shifted to a higher level of $\overline{T}b$ without changes of the inter-threshold zone. This shift is due to the mechanical compression and irritation of the hypothalamus. After decompression of this region the body temperatures and thresholds normalized. Similar changes were measured in patients with basal meningiomas. Ai are localized far from the hypothalamus, there is no mechanical alteration and no threshold shift was measured.

Hormonal findings are not suggestive for explaining these results. Basal and stimulated values of TSH were in normal range, stimulated TSH was in its lower part indicating a slight functional impairment of the thyroid-axis. However, a reduction of the thyroid function would cause a decrease of the temperatures and not an increase. As already stated, in patients with frontobasal meningiomas and completely normal thyroid-axis, similar threshold shifts were observed. Moreover, the As-n patients had lower basal and stimulated TSH levels than the As-i! Rudolf Thauer in 1939 [16] reported, that the resection of the thyroid mostly has no effect on the thermoregulatory capabilities. In monkeys, after several weeks of cold adaptation, no increase of the thyroid function was observed [11]. Much more important for cold adaptation are the catecholamines, which may act synergistically to thyroid hormones [8,11].

In acute and short term cold exposures as performed in this study (ca. 45 min. cold load) the neuronal stimulation of the primary thermogenetic mechanism in humans – muscle shivering – is of much more importance. During longer lasting exposures to cold, several studies reported an increase of TSH and several did not [9]. The reduction of the tympanic temperature of 0.5°C did not cause an elevation of TSH, nor of the growth hormone [4]. In our examinations the Tty remained stable or even increased due to peripheral vasoconstriction. In human newborns acute cold exposure activated the pituitary-thyroid-axis; this was not observed in adults [10]. Pilot measurements of TSH and thyroid hormones in our patients during cold and heat load revealed no changes. Taking this data from literature and our own findings into consideration it cannot be assumed that hormonal disturbances due to the pituitary adenomas are responsible for the thermoregulatory modifications.

The hypothalamic catecholaminergic- and serotonergic system is highly suggestive for being involved and responsible for the threshold shifts. Microinjections of noradrenaline into the hypothalamus of guinea pigs resulted in an elevation of thermogenetic thresholds [6,18]. Serotonine had an opposite effect [19]. These central amines act on interneurons of the thermoregulatory network which are responsible for the setting of the thermoregulatory thresholds. In the anterior hypothalamus the dominating temperature sensitive neurons are warmreceptors. Only 2–10% of the thermosensitive neurons are coldreceptors [7,12,17]. The mass effect of a tumor would therefore predominantly impair warmreceptors. In this situation the cold input to the regulating system, especially from the skin, would be increased facilitating cold defence reactions. In hot environments the responsiveness of hypothalamic warmsensors is reduced which results in a delayed heat loss reaction. Both mechanisms, alteration of amine responsive interneurons and central warmreceptors will produce the same modification of thermoregulatory characteristics: an elevation of threshold temperatures and of the "set-point". The accuracy of regulation is not altered because the inter-threshold zone is not widened. However, the velocity of cold- and heat-defence reactions is reduced.

Acknowledgement

Supported by DFG Grant Be-1015/1–3.

References

1. Behr R, Hildebrandt G, Koca M, Brück K (1991) Modifications of thermoregulation in patients with suprasellar pituitary adenomas. Brain 114: 697–708
2. Behr R, Hildebrandt G, Klug N, Brück K (1991) Thermoregulation in patients with skull base tumors. In: Samii M (ed) Surgery of the sella and paranasal sinuses. Springer, Berlin Heidelberg New York Tokyo, pp 395–406
3. Behr R, Dietrich Ch, Brück K (1994) Alterations of heat dissipation by diabetes insipidus in humans. In: Zeisberger E, Schönbaum E, Lomax P (eds) Thermal balance in health and disease. Birkhaeuser, Basel, pp 267–276
4. Berg GB, Utiger RD, Schalch DS, Reichlin S (1966) Effect of central cooling in man on pituitary thyroid function and growth hormone secretion. J Appl Physiol 21: 1791–1794
5. Brück K, Zeisberger E (1978) Significance and possible central mechanisms of thermoregulatory threshold deviations in thermal adaptation. In: Wang LCH, Hudson LW (eds) Strategies in cold: natural torpidity and thermogenesis. Academic Press, London, pp 655–694

6. Brück K, Zeisberger E (1987) Adaptive changes in thermoregulation and their neuropharmacological basis. Pharmac Ther 35: 163–215

7. Boulant JA, Dean JB (1986) Temperature receptors in the central nervous system. Ann Rev Physiol 48: 639–654

8. Christensen NJ (1972) Increased levels of plasma noradrenaline in hypothyroidism. L Endocrin Metab 35: 359–363

9. Fregly MJ (1990) Activity of the hypothalamic-pituitary-thyroid axis during exposures to cold. In: Schönbaum E, Lomax P (eds) Thermoregulation physiology and biochemistry. Intern Encyclop Pharmacol Therapeutics. Pergamon, pp 437–494

10. Frohman LA (1987) Diseases of the anterior pituitary. In: Felig Ph, Baxter JD, Broadus AE, Frohman LA (eds) Endocrinology and metabolism. McGraw Hill, pp 259

11. Gale CC (1973) Neuroendocrine aspects of thermoregulation. Ann Rev Physiol: 391–430

12. Hensel H (1981) Thermoreception and temperature regulation. Monogr Physiolog Soc 38

13. Hildebrandt G (1989) Neuroendokrinologische Funktionsstörungen bei intrinsischen und extrinsischen Prozessen des hypothalamo-hypophysären Systems: Eine Untersuchung aus neurochirurgischer Sicht. Habilitationsschrift, Fachbereich Humanmedizin, Justus- Liebig Universität Giessen

14. Lausberg G (1972) Zentrale Störungen der Temperaturregulation. Acta Neurochir (Wien) [Suppl 19]

15. Lees PD, Pickard JD (1987) Hyperprolactinemia, intrasellar pituitary tissue pressure, and the pituitary stalk compression syndrome. J Neurosurg 192–196

16. Thauer R (1939) Der Mechanismus der Wärmeregulation. Ergebnisse der Physiologie 41: 607–805

17. Watanabe T, Morimoto A, Murakami N (1986) Effect of amine on temperature responsive neurons in slice preparations of the rat brain stem. Am J Physiol 250: R553–R559

18. Zeisberger E, Brück K (1971) Effect of intrahypothalamic noradrenaline injections on the threshold temperatures for shivering and non-shivering thermogenesis. J Physiol (Paris) 63: 464–467

19. Zeisberger E, Wissel MR (1978) Microinjections of 5-hydroxytryptamine (5-HT) into different central parts of the thermoregulatory system in newborn and adult, cold- or warm adapted guinea pigs. Pflügers Arch 377 [Suppl R31]

Correspondence: Dr. Robert Behr, Neurochirurgische Klinik und Poliklinik, Universität Würzburg, Josef-Schneider-Str. 11, D-97080 Würzburg, Federal Republic of Germany.

Acta Neurochir (1996) [Suppl] 65: 50–53

The Molecular Biology of Hormone and Growth Factor Receptors in Meningiomas

P. Black, R. Carroll, and **J. Zhang**

Neurosurgery Laboratories, Brigham and Women's Hospital, and Brain Tumor Center, Harvard Medical School, Boston, MA, U.S.A.

Summary

Expression of a number of steroid hormone and growth factor receptors is characteristic of meningiomas. This paper reviews the analysis of receptors for progesterone, estrogen, androgen and platelet derived growth factor (PDGF) in human meningioma tissue specimens.

Progesterone receptor was assessed by Northern blot analysis and immunohistochemistry in meningioma tissue specimens. Progesterone receptor mRNA was expressed in 64% of the meningiomas examined. Immunohistochemical data correlated well with the Northern blot analysis. The staining was clearly nuclear. Expression was more common in meningioma tissue from women than from men. Analysis of receptor expression in tissue culture derived from meningioma specimens demonstrated the loss of progesterone receptor within one to two passages. It was shown that the progesterone receptor mRNA expression which is present in meningiomas is functional by transfection techniques.

The estrogen receptor was undetectable by Northern blot analysis; a small amount could be detected in meningioma tissue specimens by polymerase chain reaction (PCR).

The androgen receptor was found in 67% of the specimens examined. Like the progesterone receptor, it was more common in women than in men (69% vs. 31%). The immunohistochemical data correlated well with the Northern blot analysis, with the receptor predominantly found in the nucleus. Unlike progesterone receptor, androgen receptor expression was not lost in cell culture.

The subunits for PDGF were expressed in various quantities in meningiomas. Only the PDGF β-receptor (PDGFR-β) not α-receptor, was found in meningioma tissue specimens. In contrast, the ligands PDGF A and PDGF B were expressed in all tumors. The functionality of the PDGF β-R was determined by examining the induction of the protooncogene C-*fos* by PDGF BB in meningioma cell cultures. A significant increase in C-*fos* protein was observed with the addition of PDGF BB to meningioma cultures.

Keywords: PDGF; meningiomas; cerebral neoplasms; progesterone; estrogen; androgen; receptors.

Introduction

It has been known for some time that meningiomas contain receptors for sex hormones including progesterone and estrogen [2,5–7, 13–16,22]. The range of expression of these receptors has not been explored extensively. New molecular biological techniques have generally not been used to investigate the expression of these receptors in meningiomas. This paper summarizes several years of work by our laboratory evaluating the expression of the mRNA and protein for the receptors for progesterone, estrogen, androgen, and platelet-derived growth factor in human meningiomas.

Materials and Methods

Tissue Samples/Cell Culture

For Northern blot analysis, tissues were collected at the time of craniotomy for tumor resection and were immediately snap frozen in liquid nitrogen and subsequently stored in liquid nitrogen. Non-neoplastic tissue was obtained from patients undergoing temporal lobectomy for uncontrollable seizures, as a comparison. Each sample was taken from a specimen which was used by the neuropathologist for diagnosis. These tumors were reviewed by Dr. Matthew Frosch (Brigham and Women's Hospital). MCF-7 breast carcinoma cells were obtained from the American Culture Collection and maintained in DMEM-10% fetal bovine serum (FBS). These were used as a positive control on Northern blots and PCR for estrogen and androgen receptor. T-47D cells were obtained from the American Culture Collection and maintained in RPMI 1640, 10% FBS and 0.2 U IU insulin/ml and were used as a positive control on Northern blots and PCR for progesterone receptor mRNA expression.

RNA Isolation and Northern Blot Analysis

Total RNA was isolated by the method of Chirgwin [9]. Tissue samples were placed in 4 M guanidinium isothiocyanate and then homogenized with the use of a Polytron until they were totally disrupted. For cells, media was aspirated and cells were washed twice in ice cold phosphate buffered saline and 3.3 ml of guanidinium isothiocyanate was added to each flask. The cells were scraped into the guanidinium isothiocyanate and DNA was sheared by passage through a 21 G needle. After centrifugation for ten minutes at 3,000 rpm, 20°C (Beckman RT 6000) the supernatant was layered over 5.7 M cesium chloride and centrifuged in a Beckman ultracentrifuge in a SW 50.1 rotor at a speed of 38,000 rpm, 22°C for 16 h. The RNA pellet was dissolved in 0.3 M sterile sodium acetate and the RNA was ethanol precipitated. Twenty micrograms (A_{260}) of total RNA for each sample was subjected to electrophoresis and diffusion blotted onto Duralon nylon membrane (Strategene, La Jolla, CA) [23]. The RNA was cross-linked to the Duralon using ultraviolet light (Stratalinker, Stratagene, CA). Blots were prehybridized for two hours At 42° (50% formamide, 5X SSC (0.15 M NaCl, 0.015 M sodium citrate, Ph 7), 10X Denhardt's solution, 50 mM NaPO$_4$, 1% SDS, 10 mg/ml Sigma free acid) and hybridized overnight at 42°C (50% formamide, 5X SSC, IX Denhardt's solution, 20 mM NAPO$_4$, 0.5% SDS, 5% dextran sulfate, 20 g/ml Sigma free acid) with 10^6 cpm/ml of ^{32}P labeled cDNA probe. Northern blots were sequentially hybridized with the specific steroid receptor cDNA probe and then β-actin. Blots were washed and then subjected to autoradiography.

The blots were subsequently probed with progesterone receptor DNA kindly provided by Dr. Bert O'Malley and β-actin kindly provided by Dr. Larry Kedes. Estrogen receptor DNA was provided by Dr. Pierre Chambon. Androgen receptor cDNA was provided by Dr. Elizabeth Wilson. PDGFA and PDGFB was provided by Tucker Collins, PDGFB receptor by ATCC; PDGF α-receptor by Daniel Pope-Bowen. Blots were washed and subjected to autoradiography, and band densities were determined by laser densitometry (Molecular Dynamics, Sunnyvale, CA). The results were expressed as arbitrary densitometry units (ADU).

Results

Estrogen Receptor

Estrogen receptor was not detectable by Northern blot analysis in any of the tumors tested. PCR was able to detect estrogen receptor mRNA expression in a small proportion of meningiomas.

Progesterone Receptor [6]

By Northern blot techniques, progesterone receptor mRNA was detected in 64% of tumor samples tested. Table 1 provides the details of expression of mRNA compared with immunohistochemical expression. Immunohistochemical data correlated well with Northern blot data, demonstrating expression of the protein in the nucleus. By transient transfection of a construct containing the progesterone responsive element linked to chloramphenicol acetyl transferase (CAT) it was demonstrated that the PR is functional [8]. Moreover,

Table 1. *Progesterone Receptor Profiles in Meningiomas*

Case	Age	Sex	mRNA levels	Immunocytochemistry	Histologic type
1	64	F	++++	++++	syncytial with few psammona bodies
2	54	M	++++		transitional/syncytial
3	27	F	+++	++++	syncytial
4	43	F	+++		transitional
5	48	F	+++	++++	syncytial with rare fibroblastic areas
6	67	F	+++	++++	syncytial with rare fibroblastic areas
7	69	F	+++	++++	transitional
8	63	M	+++	++	syncytial with rare fibroblastic areas
9	27	F	++		transitional
10	39	F	++		transitional
11	39	F	++		syncytial/transitional
12	63	F	++		transitional
13	65	F	++		transitional with psammoma bodies
14	61	M	++		transitional
15	26	F	+		transitional
16	35	F	+		syncytial
17	54	F	+		fibroblastic with few psammoma bodies
18	55	F	+		fibroblastic/transitional
19	56	F	+		transitional with few psammoma bodies
20	86	F	+		syncytial with rare fibroblastic areas
21	70	M	+		syncytial/transitional with bone invasion and mitosis
22	56	F	–	–	fibroblastic with minor transitional
23	65	F	–		transitional with few psammoma bodies
24	67	F	–	+/–	transitional
25	78	F	–		syncytial
26	78	F	–	–	transitional
27	28	M	–	–	malignant/syncytial
28	45	M	–		hemanogiopericytoma
29	54	M	–		syncytial; areas of sclerosis
30	63	M	–	–	syncytial/transitional
31	64	M	–		syncytial with areas of transitional
32	65	M	–		fibroblastic/transitional
33	71	M	–		

See ref. [6]

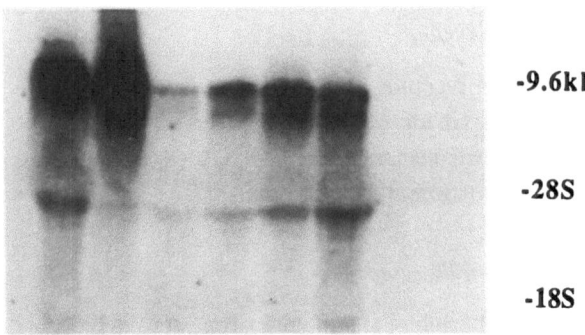

Androgen Receptor

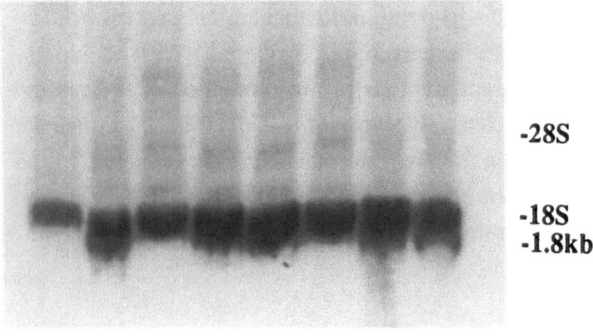

ß-Actin

Fig. 1. Expression of androgen receptor mRNA in meningioma tissue [8]

the receptor appears to be lost quickly in culture, an important point if cell culture is to be used to assess its potential usefulness therapeutically (Black *et al.*, submitted).

Androgen Receptor [8]

The androgen receptor was expressed in 67% of tumors (Fig. 1). Like the progesterone receptor, it is expressed more commonly in women than in men, with 69% of women and 31% of men expressing it. Immunohistochemical data revealed the expression of the androgen receptor in the nucleus suggesting that it is in a location to be functional.

PDGF Receptors [4]

PDGF-A and PDGF-B were expressed in all tumors in various amounts. All but one tumor expressed PDGFβ-R and no meningiomas in this series expressed PDGF αR. The functionality of the PDGFβ-R was assessed by examining the induction of the proto-oncogene C-*fos* by PDGF-BB in meningioma cul-

tures. A significant increase in C-*fos* protein was observed 3 hours after PDGF-BB addition [4].

Discussion

Meningiomas are intriguing tumors because of the variety of hormone receptors they express. This paper reviews our laboratory's experience with the molecular analysis of progesterone, estrogen, androgen, and PDGF receptors.

It is clear that progesterone receptor is expressed in meningiomas, more commonly in women than in men. The size of the mRNA species are similar to those found in T47D cells, a breast carcinoma cell line. The receptor has been shown to be functional by transient transfection experiments which utilize the progesterone-responsive element linked to a CAT reporter gene.

Attempts to demonstrate stimulation or inhibition by progesterone have not been successful in our laboratory, a finding which supports other workers who have attempted this [1,2,10,12,17,18,20]. We found that expression of progesterone receptor mRNA is lost within 1–3 passages in culture in some lines, making stimulation through that receptor unlikely to cause a growth change in culture.

Androgen receptor has previously been found by immunohistochemistry but the mRNA for it is also clearly expressed in meningiomas.

The expression of PDGF receptors is intriguing from the point of view of meningioma pathogenesis. Both A and B subunits of PDGF are found in meningioma tissue but only the receptor which responds to PDGF-BB is found there.

References

1. Adams EF, Schrell UMH, Fahlbusch R (1992) Hormonal dependence of human meningiomas Part II. In vitro effect of steroids, bromocriptine, and epidermal growth factor on the growth of meningiomas. J Neurosurg
2. Blaauw G, Blankenstein MA, Lamberts SW (1986) Sex steroid receptors in human meningiomas. Acta Neurochir (Wien) 79: 42–47
3. Black P (1993) Meningiomas. Neurosurgery 32: 643–657
4. Black P McL, Carroll R, Glowacka D, Riley K, Dashner K (1994) Platelet-derived growth factor expression and stimulation in human meningiomas. J Neurosurg 81: 388–393
5. Blankenstein MA, Blaauw G, Lamberts SW, *et al.* (1983) Presence of progesterone receptors and absence of estrogen receptors in human intracranial meningioma cytosols. Eur J Cancer Clin Oncol 19: 365–370
6. Carroll R, Glowacka R, Dashner K, Black P (1993) Progesterone receptor expression in human meningiomas. Cancer Research 53: 1312–1316

7. Carroll RS, Zhang J, Dashner K, Sar M, Wilson WM, Black P McL (1995) Androgen receptor expression in meningiomas. J Neurosurg 82: 453–460

8. Caroll RS, Zhang J, Dashner K, Black PM (1995) Progesterone and glucorticoid receptor activation in meningiomas. Neurosurgery 37: 92–97

9. Chirgwin JA, Przybyla AE, MacDonald RJ (1979) Isolation of biologically active ribonucleic acid from sources enriched with ribonuclease. Biochemistry 18: 5294–5301

10. Grunberg SM, Daniels AM, Muensch H, et al. (1987) Correlation of meningioma hormone receptor status with hor-mone sensitivity in a tumor stem-cell assay. J Neurosurg 66: 405–408

11. Grunberg SM, Weiss MH, Spitz IM, et al. (1991) Treatment of unresectable meningiomas with the antiprogesterone agent mifepristone. J Neurosurg 74: 861–866

12. Jay JR, Mac Laughlin DT, Riley KR, et al. (1985) Modulation of meningioma cell growth by sex steroid hormones in vitro. J Neurosurg 62: 757–762

13. Lesch KP, Engl HG, Gross S (1987) Androgen receptor binding activity in meningiomas. Surg Neurol 28: 176–180

14. Lesch KP, Engl HG, Schott W, et al. (1987) Immunoreactive estrogen receptor protein in meningiomas: comparison with the androgen receptor and progesterone receptor binding activity. Zentralbl Neurochir 48: 124–134

15. Lesch KP, Fahlbusch R (1986) Simultaneous estradiol and progesterone receptor analysis in meningiomas. Surg Neurol 26: 257–263

16. Lesch KP, Schott W, Engl HG, et al. (1987) Gonadal steroid receptors in meningiomas. J Neurol 234: 328–333

17. Markwalder TM, Gerber HA, Waelti E, et al. (1988) Hormonotherapy of meningiomas with medroxyprogesterone acetate. Immunohistochemical demonstration of the effect of medroxyprogesterone acetate on growth fractions of meningioma cells using the monoclonal antibody Ki67. Surg Neurol 30: 97–101

18. Markwalder TM, Markwalder RV, Zava DT (1984) Estrogen and progestin receptors in meningiomas: clinicopathological correlations. Clin Neuropharmacol 7: 368–374

19. Markwalder TM, Waelti E, Konig MP (1987) Endocrine manipulation of meningiomas with medroxyprogesterone acetate. Effect of MPA on receptor status of meningioma cytosols. Surg Neurol 28: 3–9

20. Olson JJ, Beck DW, Schlect J, et al. (1986) Hormonal manipulation of meningiomas in vitro. J Neurosurg 65: 99–107

21. Olson JJ, Beck DW, Schlect J, et al. (1987) Effect of the antiprogesterone RU486 on meningioma implanted into nude mice. J Neurosurg 66: 584–587

22. Schrell UMH, Adams EF, Fahlbusch R, et al. (1990) Hormonal dependence of cerebral meningiomas. Part 1: Female sex steroid receptors and their significance as specific markers for adjuvant medical therapy. J Neurosurg 73: 743–9

23. Schrell UMH, Fahlbusch R (1991) Hormonal manipulation of cerebral meningiomas. In: Al-Mefty O (1991) Meningiomas. Raven, New York

Correspondence: Dr. Peter Black, Neurosurgery Laboratories, Brigham and Women's Hospital, 75 Francis St., Boston, MA 02115, U.S.A.

Acta Neurochir (1996) [Suppl] 65: 54–57
© Springer-Verlag 1996

Hormonal Dependency of Cerebral Meningiomas

U.M.H. Schrell, P. Nomikos, Th. Schrauzer, M. Anders, R. Marschalek[1], E.F. Adams, and R. Fahlbusch

Departments of Neurosurgery and Genetics[1], University of Erlangen-Nürnberg, Erlangen, Federal Republic of Germany

Summary

Though meningiomas are benign intracranial tumors, a minor group invades the skull base and the connective tissue of the sinus cavernous inducing neurological deficits. These patients can not be cured by surgical treatment. Therefore, the development of an adjuvant medical therapy has been the goal during the last decade. We report here on different medical concepts which are based on steroids, amines, growth factor antagonists and cytokines. In addition, our data give evidence that the growth of intracranial meningiomas is under multifactorial proliferative control.

Keywords: Meningioma; steroids; amines; growth factors; cytokines.

Introduction

During the last decade considerable attention has been given to adjuvant medical treatment of recurrent cerebral meningiomas because of their invasiveness into the skull base and cavernous sinus or multiple meningiomatosis. During the past 13 years various concepts of influencing meningioma cell growth have been developed with the intention of controlling the proliferation of this tumor [19]. Steroidal growth control has been the primary concept and is still the most controversial one.

Steroid Dependency

Using a dextran-coated charcoal assay (DCCA), a solid-phase enzyme immunoassay (ELISA), immunohistochemistry and in situ hybridisation with a synthetic oligonucleotide probe complementary to a relevant fraction of the estrogen receptor mRNA, we and other groups have provided evidence that the estrogen receptor is generally absent [13,14,18]. The progesterone receptor was regularly found in the cytoplasm of meningioma cells, although the active receptor complex was found in the nucleus in a minor percentage raising the question whether only a part of the progesterone receptors are transcriptionally active in cerebral meningiomas. In vitro studies with cultured meningioma cells supported this findings [2]. However, recent data gave evidence for expression and translation of the progesterone receptor mRNA in meningioma tissue [6]. Clinical trials with the antiprogesterone RU 486 led to a significant shrinkage of meningioma growth only in a small group of patients with inoperable and recurrent meningiomas [9,15].

Amine Dependency

Treating meningioma cell cultures with dopamine agonists, i.e. the dopamine D2 agonist bromocriptine and selective dopamine-D1 agonists (SKF 82958 and SKF 38393), was found to decrease meningioma cell growth up to 50% [20]. DNA flow-cytometric-analysis revealed an increase in G2+M- and/or S-phase in the treated cells. With respect to the reduced proliferation rate, these data provided evidence for an arrest of meningioma cells in these cell cycle phases. The finding that the dopaminergic agents are only effective when applied in μM concentrations, indicated the presence of dopamine-D1 receptors in meningiomas. Saturation and Scatchard analysis [1] in 45 cerebral meningiomas revealed the presence of high affinity binding sites in 33 meningiomas with a dissociation constant (Kd) of 369 pM ± 169 SD and a receptor density (Bmax) of 31.9 fmol/mg membrane pellet protein ± 12.5 SD [23]. Because we failed to find the dopamine D2 receptor in meningioma tissue we have to conclude that the antiproliferative effect of bromocryptine is independent of this receptor type.

Recently, we have found that a second aminergic receptor types, the serotonin 5-HT1C and 5-HT2 receptors, are present in cerebral meningiomas [24]. Saturation curves and the linearity of the Scatchard analysis indicated that the serotonin antagonist [125I]ketanserin binds to high affinity binding sites in 25 of 28 human cerebral meningiomas. The mean (± standard deviation) dissociation rate constant (Kd) was 483.44 ± 150.07 pM with a density (Bmax) of 345.96 ± 74.82 fmol/mg membrane protein among 25 meningioma specimens. Because [125I]ketanserin is a 5-HT2 and 5-HT1C receptor antagonist, it is reasonable that both receptor types may be expressed. Though the rank order of the dissociation curves argues in favour of a 5-HT1C receptor binding site, the presence of the 5-HT2 receptor type cannot be excluded by this profile. Extraction of mRNA of low passage meningioma cell cultures, RT-PCR, direct sequencing and cloning of the PCR-product proves the expression of both receptor types, the 50-HT1C [10,16] and 5-HT2 [7] receptors, in cultured meningioma tissue.

To investigate whether serotonin modulates meningioma cell growth, we used the [3H]-thymidine incorporation into DNA. Serotonin in concentrations of 10-9 to 10-5M showed a dose dependent [3H]-thymidine uptake up to 100% over control and proved serotonin to be a potent mitogen in cultured meningioma cells and supports the concept that biogenic amines are involved in meningioma cell growth.

Growth Factor Dependency and Suramin

Growth of human cerebral meningiomas is under control of various growth factors, i.e. EGF [30,31], PDGF-BB [17], IGF I and II [4] and acidic and basic FGF [27,28]. The latter three have been shown to form autocrine growth loops in meningiomatous tissue which may be the main factor for uncontrolled growth of these tumors [3–5,11,17,25,26]. Suramin is known to prevent binding of a variety of growth factors to their receptors in mammalian tissue thus abolishing para- and/or autocrine mediated cell growth. We therefore tested the effect of suramin on the proliferation of cultured human meningioma cells [21].

Suramin (10^{-5} to 10^{-4} M) significantly inhibited the growth of meningioma cells in culture. The maximum effect observed was with the high dose (10^{-4} M) which resulted in a 40 to 50% reduction of cellular proliferation. In studies using DNA flow-cytometry, suramin inhibited meningioma cell proliferation by arresting cells in the S- and G_2M phase of cell cycle. Growth factor (EGF, IGF I and PDGF-BB) induced cell proliferation was completely abolished when suramin (10^{-4} M) was applied to meningioma cells. Western blot showed that the intracellular PDGF-BB content of meningioma cells was significantly reduced after treating the cells with suramin (10^{-4} M). Binding of iodinated growth factors (i.e. [125I]EGF, [125I]IGF I and [125I]PDGF-BB) to their receptor sites in meningioma membrane fractions was prevented in a dose dependent manner by suramin. Lowering the intracellular PDGF content and preventing extracellular growth factor receptor binding, it is evident that suramin disrupts autocrine loops and paracrine growth stimulation in meningioma tissue.

These data showed clearly that growth of cerebral meningiomas in culture is strongly inhibited by suramin in concentrations of 10^{-4} M. Cerebral meningiomas grow under control of various growth factors and suramin acts as a scavenger neutralizing exogenous growth factors. So it can interrupt autocrine loops and paracrine stimulation of human meningioma cell. This favours suramin as a therapeutic option for controlling meningioma proliferation in patients with inoperable and recurrent high grade meningiomas.

Cytokine Dependency

The cytokines leukaemia inhibitory factor (LIF) [8] and interleukin-6 (IL-6) [12,32], are produced by various tumorous and non-tumorous tissues. They are known to be involved in a variety of biological processes such as the induction of acute-phase proteins, the induction of cell differentiation and inhibition of proliferation in some cell types. As the proliferation of human cerebral meningiomas is known to be under control of various growth factors, we set out to investigate the expression of cytokines, especially LIF and IL-6, in cerebral meningiomas [22,29].

As determined by ELISA, low passage human meningioma cells secrete LIF in a concentration of 40 to 200 pg per 10^6 cells and IL-6 in a concentration of 5 to 100 ng per 10^6 cells. The amount of LIF and IL-6 secreted in culture correlated positively with the increase of cell numbers. Western blot showed a molecular weight of 30 kD for IL-6 and 48 kD for LIF. RT-PCR, using mRNA from cultured meningioma tissue and four specific primer pairs, overspanning an intron, revealed the expression of LIF, IL-6 and their receptors (LIF-R, IL-6-R). Direct dideoxy sequencing of these RT-PCR products showed the expected four cDNA sequences.

Our data gave evidence for the presence of LIF- and IL-6 protein in human cerebral meningioma tissue. The expression of LIF and IL-6 mRNA as well as the expression of mRNA for the LIF- and IL-6-receptor in the same tissue gave evidence that these cytokines form autocrine loops in human cerebral meningiomas.

We have presented different concepts and summarized various factors that influence the growth of cerebral meningiomas. At present it cannot be foreseen which system is the dominant one. This multitude of regulatory influences imply that the growth of meningiomas is under multifactorial proliferative control. This has to be taken into account when developing adjunctive drug therapies.

References

1. Abramson RD, Nies AS, Gerber JG, Molinoff PB (1987) Evaluation of models for analysis of radioligand binding data. Mol Pharmacol 31: 103
2. Adams EF, Schrell UMH, Fahlbusch R, Thierauf P (1990) Hormonal dependency of cerebral meningiomas. Part 2: In vitro effect of steroids, bromocriptine, and epidermal growth factor on growth of meningiomas. J Neurosurg 73: 750
3. Adams EF, Todo T, Schrell UMH, Thierauf P, White MC, Fahlbusch R (1991) Autocrine control of human meningioma proliferation: secretion of platelet-derived growth-factor-like molecules. Int J Cancer 49: 398
4. Antoniades HN, Galamopoulos T, Neville-Golden J, Maxwell M (1992) Expression of insulin-like growth factors I and II and their receptor mRNA in primary human astrocytomas and meningiomas; in vivo studies using in situ hybridisation and immunocytochemistry. Int J Cancer 50: 215
5. Black PM, Carroll R, Glowacka D, Riley K, Dashner K (1994) Platelet-derived growth factor expression and stimulation in human meningiomas. J Neurosurg 81: 388
6. Carroll RS, Glowacka D, Dashner K, Black PM (1993) Progesterone receptor expression in meningiomas. Cancer Res 53: 1312
7. Chen K, Yang W, Grimsby J, Shih JC (1992) The human 5-HT2 receptor is encoded by multiple intron- exon gene. Mol Brain Res 14: 20
8. Gearing DP, Gough NM, King JA, Hilton DJ, Nicola NA, Simpson RJ, Nice EC, Kelso A, Metcalf D (1987) Molecular cloning and expression of cDNA encoding a murine myeloid leukemia inhibitory factor (LIF). EMBO 6: 3995
9. Grunberg SM, Weiss MH, Spitz IM, Ahmadi J, Sadun A, Russell CA, Lucci L, Stevenson LL (1991) Treatment of unresectable meningiomas with the antiprogesterone agent mifepristone. J Neurosurg 74: 861
10. Julius D, MacDermott AB, Axel R, Jessell TM (1988) Molecular characterisation of a functional cDNA encoding the serotonin 1C receptor. Science 241: 558
11. Kaplan PL, Anderson M, Ozanne B (1982) Transforming-growth-factor(s) production enables cells to grow in the absence of serum: an autocrine system. Proc Natl Acad Sci USA 79: 485
12. Kishimoto T (1989) The biology of interleukin-6. Blood 74: 1
13. Koehorst SGA, Jacobs HM, Thijssen JHH, Blankenstein MA (1993) Wild type and alternatively spliced estrogen receptor messenger RNA in human meningioma tissue and MCF7 breast cancer cell. J Steroid Biochem Molec Biol 45: 227
14. Koehorst SGA, Jacobs HM, Thijssen JHH, Blankenstein MA (1993) Detection of an oestrogen receptor-like protein in human meningiomas by band shift assay using a synthetic oestrogen responsive element (ERF). Br J Cancer 68: 290
15. Lamberts SWJ, Tanghe HLJ, Avezaat CJJ, Braakman R, Wijngarde R, Koper JW, Jong H (1992) Mifepristone (RU 486) treatment of meningiomas. J Neurol Neurosurg Psychiatry 55: 486
16. Lubbert H, Hoffman BJ, Snutch TP, Dyke TV, Levine AJ (1987) cDNA cloning of a serotonin 5-HT1C receptor by electrophysiological assay of mRNA-injected in Xenopus oocytes. Proc Natl Acad Sci USA 84: 4332
17. Maxwell M, Galanopoulos T, Hedley-Whyte T, Black PM, Antoniades HN (1990) Human meningiomas co-express platelet-derived growth factor (PDGF) and PDGF-receptor genes and their protein products. Int J Cancer 46: 16
18. Schrell UMH, Adams EF, Fahlbusch R, Greb R, Jirikowski G, Prior R, Ramalho-Ortigao F (1990) Hormonal dependency of cerebral meningiomas. Part 1: Female sex steroid receptors and their significance as specific markers for adjuvant medical therapy. J Neurosurg 73: 743
19. Schrell UMH, Fahlbusch R, Adams EF (1994) Meningiomas and neurofibromatosis for the oncologist. Curr Opin Oncol 6: 247
20. Schrell UMH, Fahlbusch R, Adams EF, Nomikos P, Reif M (1990) Growth of cultured human cerebral meningioma is inhibited by dopaminergic agents. Presence of high affinity dopamine- D1 receptors. J Clin Endocrinol Metab 71: 1669
21. Schrell UMH, Gauer St, Kiesewetter F, Bickel A, Hren J, Adams EF, Fahlbusch R (1995) Suramin inhibits proliferation of human cerebral meningioma cells. Effects on cell growth, cell cycle phases, extracellular growth factors and on PDGF-BB autocrine growth loop. J Neurosurg: in press
22. Schrell UMH, Koch U, Schrauzer Th, Nomikos P, Fahlbusch R (1994) Secretion of leukaemia inhibitory factor and interleukin 6 and expression of their mRNAs as well as their receptor mRNAs in human cerebral meningioma tissue. Exp Clin Endocrinol 102: 188
23. Schrell UMH, Nomikos P, Fahlbusch R (1992) Presence of dopamine D1 receptors and absence of dopamine D2 receptors in human cerebral meningioma tissue. J Neurosurg 77: 288
24. Schrell UMH, Nomikos P, Marschalek R, Haschke, Schrauzer T, Anders M, Adams E, Fahlbusch R (1994) Expression of serotonin 5-HT1C and 5-HT2 receptors in human cerebral meningiomas. J Neurosurg: submitted
25. Sporn MB, Roberts AB (1985) Autocrine growth factors and cancer. Nature 313: 745
26. Sporn MB, Todaro GJ (1980) Autocrine secretion and malignant transformation of cells. N Engl J Med 303: 878
27. Takahashi JA, Mori H, Fukumoto M, Igarashi K, Jaye M, Oda Y, Kikuchi H, Hatanaka M (1990) Gene expression of fibroblast growth factors in human gliomas and meningiomas: demonstration of cellular or basic fibroblast growth factor mRNA and peptide in tumor tissue. Proc Natl Acad Sci USA 87: 5710
28. Takahashi JA, Suzui H, Yasuda Y, Ito N, Ohata M, Jaye M, Fukumoto M, Oda Y, Kikuchi H, Hatanaka M (1991) Gene expression of fibroblast growth factor receptor in the tissue of human gliomas and meningiomas. Biochem Biophys Res Commun 177: 1
29. Waelti ER, Markwalder R, Markwalder TM (1991) Interleukin 6 Produktion in primaeren Zellkulturen von Meningiomzellen. Schweiz Med Wschr 121: 1512
30. Weisman AS, Villemure JG, Kelly PA (1986) Regulation of DNA synthesis and growth of cells derived from primary meningiomas. Cancer Res 46: 2545

31. Westphal M, Herrmann HD (1986) Epidermal growth factor-receptors on cultured human meningioma cells. Acta Neurochir (Wien) 83: 62
32. Yamasaki K, Taga T, Hirata Y, Yawata H, Kawanishi Y, Seed B, Taniguchi T, Hirano T, Kishimotor T (1988) Cloning and expression of the human interleukin-6 (BSF-2/IFN beta 2) receptor. Science 243: 825

Correspondence: U.M.H. Schrell, M.D., Department of Neurosurgery and Genetics, University of Erlangen-Nürnberg, Schwabachanlage 6, D-91054 Erlangen, Federal Republic of Germany.

Acta Neurochir (1996) [Suppl] 65: 58–62

Surgical Treatment of Meningiomas Involving the Cavernous Sinus: Evolving Ideas Based on a Ten Year Experience

L.N. Sekhar[1], **S. Patel**[3], **M. Cusimano**[4], **D.C. Wright**[1], **C.N. Sen**[5], and **W.O. Bank**[2]

[1]Department of Neurological Surgery, [2]Department of Radiology, George Washington University Medical Center, Washington, DC, [3]Medical University of South Carolina, [4]University of Toronto, Canada, and [5]Mt. Sinai School of Medicine, New York, NY, U.S.A.

Summary

The outcomes of 114 patients with meningiomas operated at the University of Pittsburgh were analyzed. Cerebrospinal fluid leakage was the most frequent complication, observed in 25 patients (21%). Complications were more frequent in patients who had recurrent (previously operated) tumors and patients with extensive tumors. Our current analysis also indicates that patients with prior radiotherapy (usually external beam) have unacceptably high complication rates after microsurgery. Early results indicate that regrowth rates are much higher in patients with incomplete resection (20%) than those with gross total excision (5%). Of the 114 patients, 108 returned to independent living and/or their previous occupation.

Keywords: Cavernous sinus; meningioma; microsurgery.

Introduction

During the last ten years, the senior author and his colleagues have operated on 350 neoplastic and vascular lesions involving the cavernous sinus area. Among these, 114 meningiomas involving the cavernous sinus were operated while at the University of Pittsburgh (by LNS, DCW, and CNS) and 35 meningiomas were operated at George Washington University (by LNS and DCW). All of the patients operated in Pittsburgh had at least a year of follow up, and the results have been carefully analyzed (by SP and MC). This paper briefly presents the results and some of the lessons learned from the experience.

Classification

Meningiomas involving the CS are classified in two ways. The *first scheme* is based on the nature of the intracavernous ICA (CS-ICA) involvement, and the extent of CS invasion. According to this scheme, tumors which only displace the intracavernous ICA or encase it partially are classified as grades I and II; tumors which encase the CS-ICA, or encase and narrow the CS-ICA are classified as grades III and IV; and tumors with bilateral CS extension are classified as grade V. The *second scheme* of classification is based on the size and extracavernous extensions of the tumor. Thus, lesions that involve the CS and areas immediately adjacent to it and are less than 3 cm in diameter are called *confined lesions*; lesions that extend to multiple areas of the cranial base, or are more than 3 cm in diameter, are called *extensive lesions*. Approximately two thirds of all lesions operated by us were extensive.

Preoperative Evaluation

Almost all of the patients were evaluated with preoperative magnetic resonance imaging (MRI) and bone window computed tomography scans. Cerebral angiography was performed in all patients. The patients treated in Pittsburgh underwent a carotid occlusion test combined with clinical testing and Xenon blood flow studies. Embolization of the external carotid feeders to the tumors was performed.

More recently at George Washington University, we perform a carotid occlusion test only if the CS-ICA is encased by the tumor. Only clinical testing of patients during ICA occlusion is performed, combined with injection of contrast into the contralateral ICA and into one vertebral artery to demonstrate the source of collateral circulation (Figs. 1–4). We do not currently perform CBF studies during carotid occlusion because of the present philosophy of reconstructing all

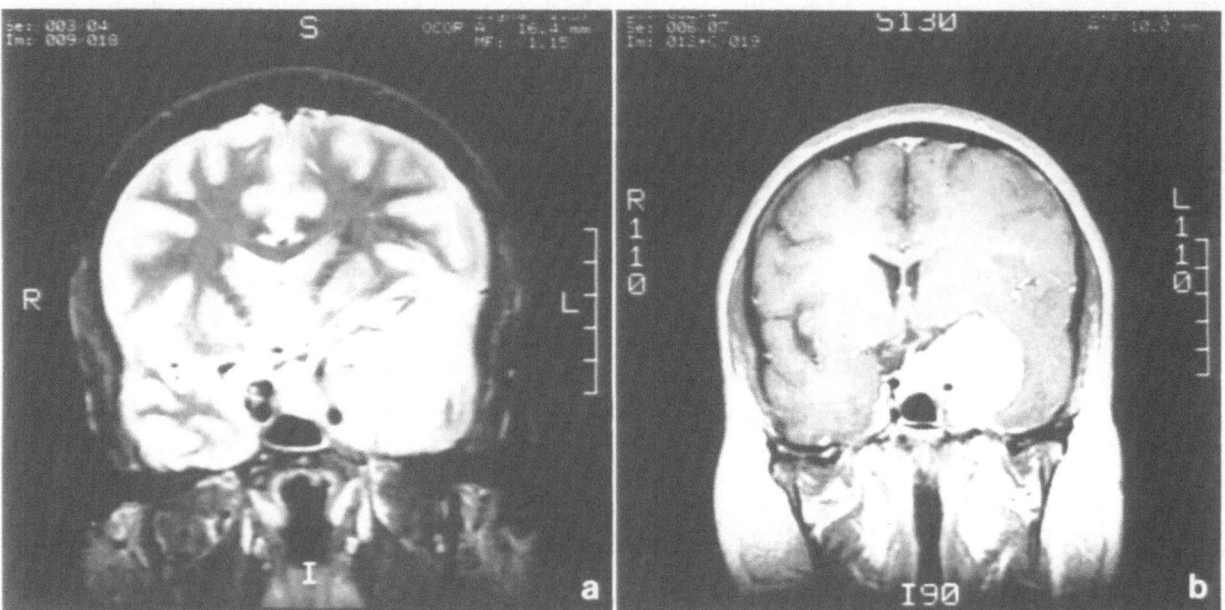

Fig. 1.(a,b) Treatment of a giant sphenocavernous meningioma in a young man who presented with mild diplopia, headaches and memory loss. A large tumor encasing the intracavernous and supraclinoid ICA is seen in (a,b). (a) Is a T2 weighted MRI scan showing considerable temporal lobe edema. (b) Is a gadolinium enhanced MRI showing tumor extent

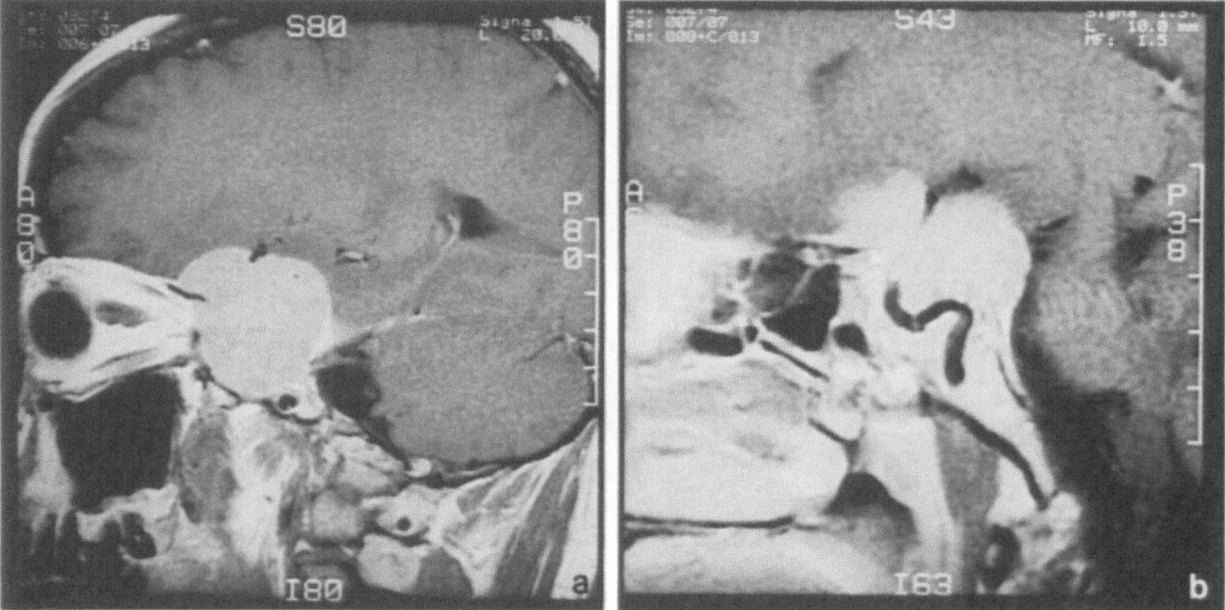

Fig. 2.(a,b) The severe encasement and some narrowing of the intracavernous and supraclinoid ICA is seen in these gadolinium enhanced MRI scans

excised ICAs regardless of the extent of collateral circulation. Because of more advanced techniques, the meningo-hypophyseal artery is nowadays routinely catheterized and embolized (by WOB) when it is feeding the tumor.

Operative Techniques

The operative techniques have evolved considerably during the period of study, especially in the last two years. In Pittsburgh, we exposed the petrous ICA for proximal control, and exposed the tumor via a frontotemporal craniotomy, and an orbitozygomatic

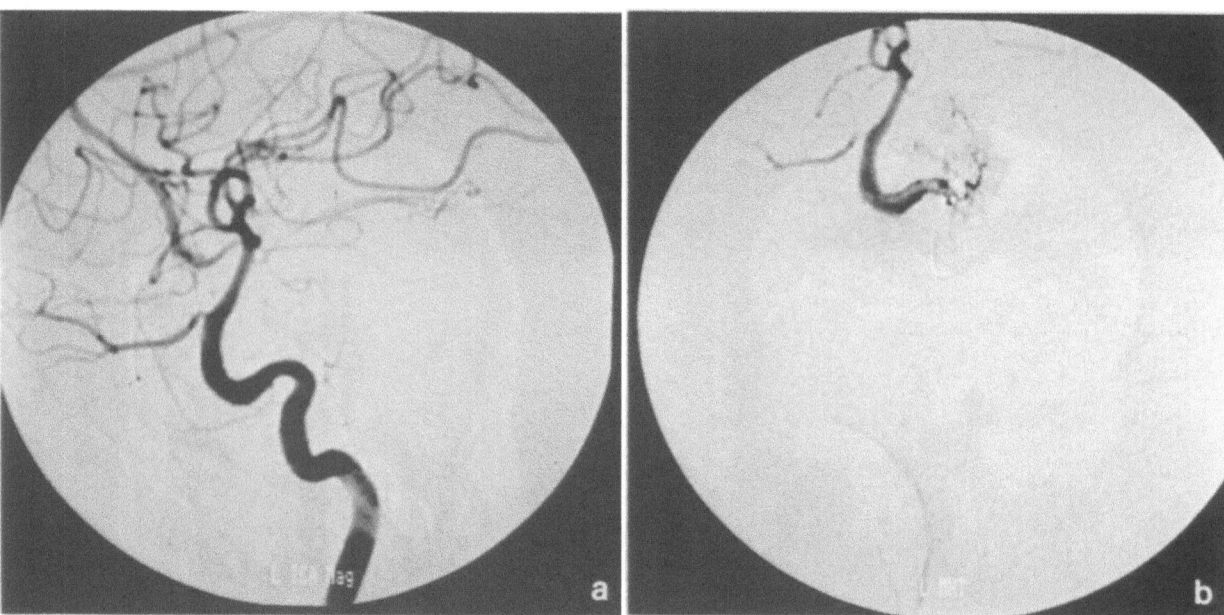

Fig. 3.(a,b) The preoperative angiogram confirms carotid artery stenosis. The meningohypophyseal artery supply to the tumor was embolized

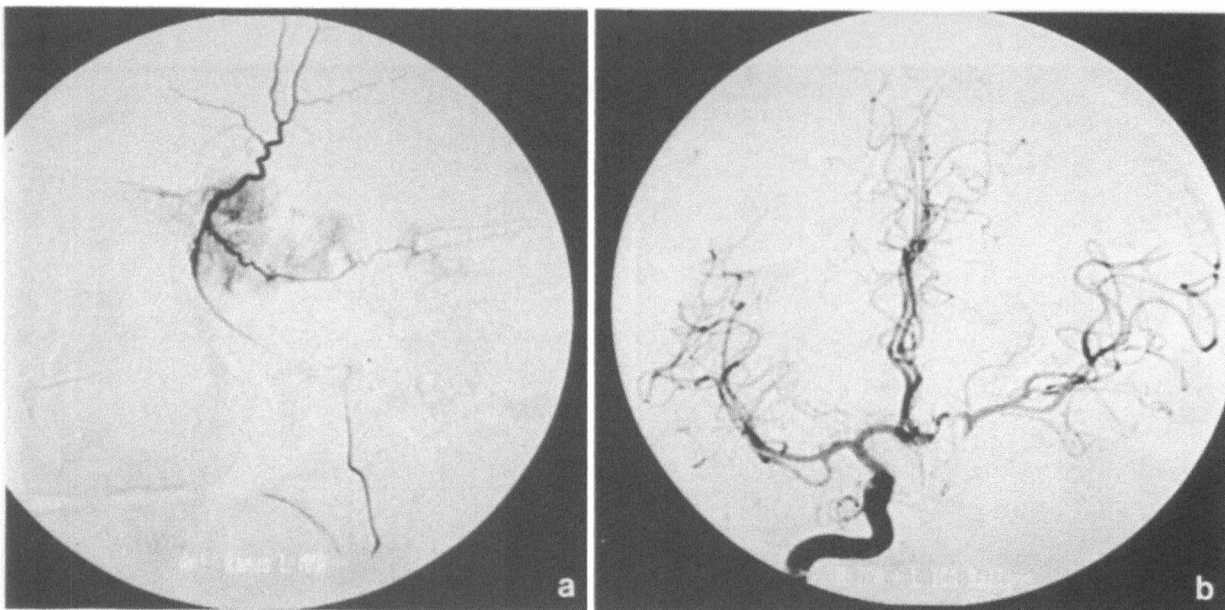

Fig. 4.(a,b) (a) Shows tumor supply through the anterior ramus of the middle meningeal artery which was embolized. (b) Reveals the right carotid arteriogram during a test occlusion of the left ICA

osteotomy. In most patients, a radical tumor excision was attempted, including the involved sphenoid bone, whenever possible. Tumor encased CS-ICA was usually excised. Patients with impaired collateral circulation were usually revascularized by means of a petrous to supraclinoid ICA graft. This was also performed in some patients with adequate collateral circulation. Any cranial nerves which were injured during operation or invaded by tumor were reconstructed with grafts. Radiosurgery was used in a few patients, as an adjunct to therapy.

At George Washington University, we prefer the exposure of the cervical ICA for proximal control, in order to avoid the division of the greater superficial petrosal nerve (GSPN). It was found in earlier patients that the division of the GSPN causes a dry eye, and when this is combined with a dry cornea patients are more liable to develop corneal ulceration. We presently prefer to revascularize all patients in whom the CS-ICA is occluded, with the aid of a long saphenous vein graft from the cervical ICA or ECA, to the M2 segment of the middle cerebral artery (Figs. 5 and 6). Because of the routine use of intraoperative angiography and immediate correction of any problems which are discovered, the patency rate has improved to 100%. This type of graft has the advantages of a much shorter ischemia time, and a graft away from the tumor site.

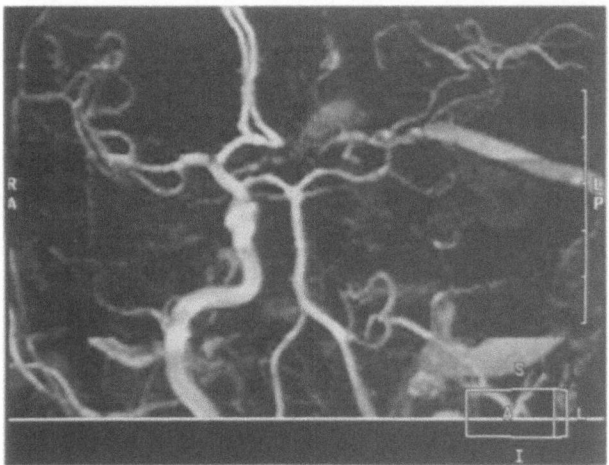

Fig. 5. The tumor was excised totally, but in three operations. During the first procedure, the extracavernous tumor, encasing the ICA, was removed. During the second procedure, a saphenous vein graft was placed from the cervical ICA to the M2 segment of the middle cerebral artery. A follow up MR angiogram shows the patent vein graft

The decision to revascularize all patients after ICA excision is based on review of data from patients with good collateral circulation who underwent elective occlusion of the artery in Pittsburgh. We discovered that there was an unacceptably high stroke rate in these patients, because of thrombo-embolic complications [1].

We have become more selective in attempting radical removal. Some tumor is left around cranial nerves when it is densely adherent or invasive, in patients who have good or excellent extraocular muscle (EOM) function. A more radical resection (with CN reconstruction, if necessary) is still performed in young patients, and patients who have only fair EOM function. Tumor residues in patients are observed with serial imaging studies, or treated with radiosurgery. The incidence of CSF leakage has been greatly reduced because we do not remove the medial sphenoid bone as aggressively. Finally, we operate on very few previously irradiated patients, because of the increased complication rates in this group.

Results

The 114 patients operated at Pittsburgh were analyzed in detail (by SP and MC). Six patients in this group suffered cerebral infarction, only two of them being related to the intracavernous ICA (one vein graft occlusion, one related to preoperative ICA occlusion

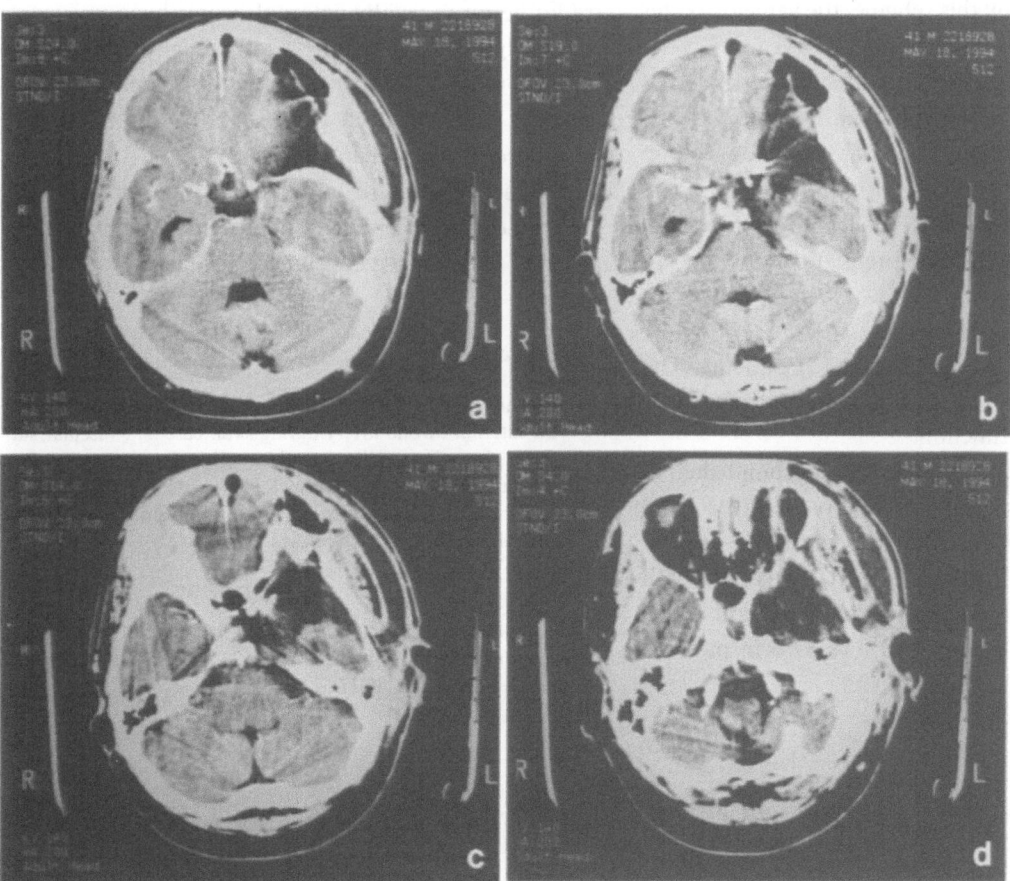

Fig. 6.(a–d) During the third operation, the remaining portion of the intracavernous tumor was excised. The enhanced CT scan shows the absence of tumor and the vein graft. Pathological examination of the intracavernous ICA showed invasion of the tunica media and tunica adventitia by the tumor. The patient suffered palsies of CNs III and VI postoperatively, from which he is recovering. He was able to return to work as an engineer, three months after the first operation

and subsequent embolism). Cerebrospinal fluid leakage was the most frequent complication, being observed in 25 patients (21%). Complications were more frequent in patients with recurrent (previously operated) tumors, and patients with extensive tumors.

Despite the frequent occurrence of complications, the eventual outcome was gratifying. One patient died of pulmonary embolism (0.8%), three suffered severe disability, and two suffered mild or moderate disability (4%). The remainder returned to independent living and/or to their prior occupation.

Extraocular muscle function was assessed according to a scheme developed by two ophthalmologists [2]. *Excellent* binocular vision consisted of what every individual uses in daily life, i.e. binocular vision present in primary and reading position, and for 20° of gaze in different directions. Beyond this, the normal person turns the head rather than the eyes. *Good* consisted of binocular vision in the reading and primary positions. *Fair* indicated moderate blepharoptosis, and/or binocular vision achievable only by head tilt. *Poor* indicated a near complete blepharoptosis, or ophthalmoplegia. Assessed by this scheme, the same or better postoperative EOM function was observed in 43% of patients who had excellent preop function, 38% of patients with good preop function, 50% of patients with fair preop function, and 57% with poor preop function. Excellent, good, or fair function was observed in 74% of patients with excellent preop function, and 63% of patients with good preop function. Postoperative binocular function was worse in patients with previous surgery or radiotherapy, and in grade III or IV patients compared to grade I and II patients. The placement of an ICA vein graft did not influence postoperative binocular function.

Overall, a gross total resection was accomplished in 78% of all patients. The probability of gross total resection was less in patients with grade III and IV tumors (73% vs 90% in grades I and II), in patients with extensive tumors (76% vs 82% in confined lesions), and patients with prior therapy (68% vs 87% in patients without).

The follow up period (median 3.9 years) is too short to assess recurrence of these meningiomas. However, early results indicate that the regrowth rates were much higher in patients with incomplete resection (20%) than those with gross total excision (5%). This was despite the use of adjuvant therapy in patients with incomplete resection.

Conclusions

Microsurgery for lesions within the CS has evolved from phases of skepticism and enthusiasm to one of careful evaluation of results, and adjustments based on the results. Further and careful analysis of patient outcome and costs of treatment will influence the nature of therapy in future. Rehabilitation of patients, especially of their eye function, is an important component of therapy.

The long-term results of radiosurgery need to be evaluated critically so that appropriate conclusions can be drawn regarding its efficacy and complications. Our current analysis indicates that patients with prior radiotherapy (usually external beam) have unacceptably high complication rates after microsurgery. Therefore, if a number of patients treated by radiosurgery will later need microsurgery, this factor will also need to be evaluated.

Further efforts to understand the biology and genesis of meningiomas, and to treat them by nonsurgical means must continue vigorously.

References

1. Sekhar LN, Patel ST (1993) Editorial – permanent occlusion of the internal carotid artery during skull base and vascular surgery: is it really safe? Am J Otol 14: 421–422
2. Biglan, AW, Sekhar LN, Cheng KP, Wright DC (1994) A protocol for measuring ophthalmologic morbidity and recovery after cranial base surgery. Skull Base Surg 4: 26–31

Correspondence: Laligam N. Sekhar, M.D., F.A.C.S., George Washington University Medical Center, 2150 Pennsylvania Avenue, NW, Washington, DC 20037, U.S.A.

Acta Neurochir (1996) [Suppl] 65: 63–65

Meningiomas Involving the Parasellar Region

M. Samii, M. Tatagiba, and **M.L. Monteiro**

Neurochirurgische Klinik, Krankenhaus Nordstadt, Hannover, Federal Republic of Germany

Summary

The authors report on 180 cases of meningiomas involving the parasellar region, which have been surgically treated between 1978 and 1993. Most of the tumors originated in the middle cranial fossa (66%). Half of the patients had visual deficits, and palsy of the eye movements was observed in 25% of the cases. Depending on the tumor origin and extension, the surgical approach was chosen. Total tumor resection varied very much, depending mostly on the tumor extension, and the pattern of growth. The overall total resection (Simpson I-II) was 57%. The postoperative results were good in 77% of the cases, and surgical mortality was 3%.

Keywords: Approaches; meningiomas; sella region; skull base.

Introduction

Meningiomas involving the sella and parasellar region originate most frequently from the tuberculum sellae, anterior clinoid or planum sphenoidale. Additionally, meningiomas arising from the anterior fossa, middle fossa and the posterior fossa may extend into the parasellar region. Visual symptoms and headaches are the most usual symptoms. Due to the complex neurovascular anatomy of the parasellar region, tumors affecting this area are difficult to resect without causing neurological deficits. Since the development of MRI and new skull base approaches, surgical treatment of meningiomas involving the parasellar region has become more feasible.

Material and Methods

Between 1978 and 1993, 180 meningiomas involving the parasellar region have been surgically treated at the Neurosurgery Department of the Nordstadt Hospital in Hannover. These 180 cases accounted for 40% of all 450 skull base meningiomas that have been operated in our Department.

All cases have been retrospectively analysed. Clinical and operative records, radiological examinations, and slides archives have been carefully investigated. Long follow-up results were not available yet in a significant number of the patients to be mentioned in this report. Thus, the postoperative results are referred to the time of patient's discharge from the hospital.

Results

There were 75% women and 25% men in this series of 180 meningiomas affecting the parasellar region. The average age was 53 years, ranging from 17 to 84 years.

The histological subtypes of the meningiomas were as follows: meningotheliomatous 76%, fibrous 4%, transitional 14%, psammomatous 3%, angioblastic and hemangiopericytic 1.5% each.

A total of 119 tumors (66%) originated in the middle cranial fossa, 38 tumors (21%) originated in the posterior fossa, and 23 tumors (13%) originated in the anterior cranial fossa. While 160 meningiomas had a single site of origin, 20 tumors showed diffuse origin through the skull base. Out of 65 cases of petro-clival meningiomas that have been operated in our Department, 33 had parasellar extension (50%) (Fig. 1). Table 1 summarizes the location of the 180 meningiomas.

Cranial nerves II to V were frequently affected. 52% of the patients had visual deficits. Due to a large number of petroclival meningiomas (33 cases) in this series of parasellar meningiomas, VIII cranial nerve affection was a common finding (22 patients).

Surgical resection was performed in all cases. The type of approach was chosen depending on the tumor origin and extension. Five approaches were most commonly used (Fig. 2): the bifrontal approach for meningiomas arising from the planum sphenoidale and olfactory groove; the frontolateral approach for

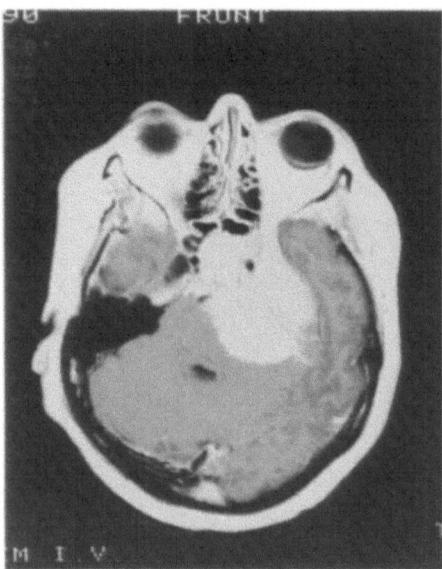

Fig. 1. Enhanced MRI scan of a petroclival meningioma showing the tumor extension into the parasellar area

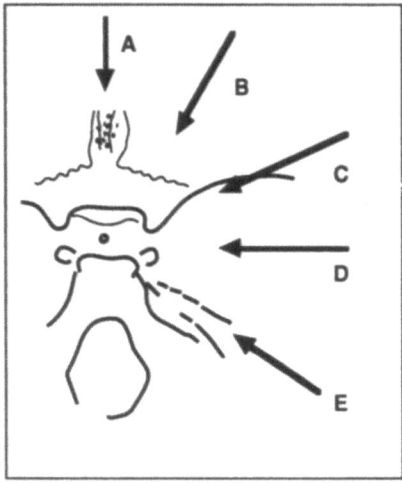

Fig. 2. Schematic drawing of the sella and parasellar region showing the most common approaches used: *A* subfrontal approach; *B* frontolateral approach; *C* pterional approach; *D* subtemporal approach; *E* presigmoid transpetrosal approach

Table 1. *Location of 180 Meningiomas with Parasellar Extension*

Origin	Number
Single origin	160
Sphenoid wing	42
Petroclival	33
Tuberculum sellae	30
Planum sphenoidale	10
Tuberculum sellae + planum sphenoidale	05
Anterior clinoid	03
Sella	08
Cavernous sinus	08
Olfactory groove	08
Orbita	04
Optic sheath	03
Meckel's cave	02
Pterygopalatine fossa	01
Cerebellopontine angle	03
Multiple origin	20
Anterior cranial fossa	03
Middle cranial fossa	09
Anterior and middle cranial fossae	01
Middle and posterior cranial fossae	04
Anterior, middle and posterior cranial fossae	03

Table 2. *Meningiomas with Parasellar Extension.* Results of tumor resection in relation to tumor origin and extension

Tumor origin	Total resection (%)	Partial resection (%)
Tuberculum sellae, olfactory groove, optic sheath, anterior cranial fossa	100	0
Planum sphenoidale	90	10
Sella	75	25
Cavernous sinus	13	87
Extended tumors (anterior, middle, and posterior fossae)	0	100
Total	57%	43%

tumor was subtotally resected. However, depending on the tumor origin and extension, complete tumor resection varied from 100% (e.g., tuberculum sellae meningioma) to 0% (e.g., diffuse growing tumor extending through the posterior and middle cranial fossae). Table 2 summarizes the results of tumor resection.

The rate of surgical mortality was 3% and the rate of surgical complication was 23%. Most frequent surgical complications were CSF leakage (8 cases), and subdural hygroma (4 cases). Transient hemiparesis was observed in 11 patients, and permanent hemiplegia was observed in one patient. Table 3 shows the new cranial nerve deficits after surgery.

At early postoperative period (time of discharge from the hospital), 77% of the patients were independent, 10% needed nursing assistance, and 10% needed hospitalization.

meningiomas arising from the tuberculum sellae; the pterional transsylvian approach for most tumors arising in the sella region, cavernous sinus, anterior clinoid, and sphenoid wing; subtemporal approach for middle fossa meningiomas; transpetrosal approach for petroclival meningiomas.

The overall total tumor resection (Simpson I-II) was 57% in this series of 180 meningiomas. In 43% the

Table 3. *Meningiomas with Parasellar Extension.* Postoperative results of cranial nerves

Cranial nerves	New deficits (%)
I	2
II	3
III	12
IV	4
V	3
VI	9
VII	7
VIII	4
IX–XII	3

Discussion

Meningiomas of the suprasellar region most commonly arise from the planum sphenoidale or tuberculum sellae [2]. However, meningiomas arising from distant areas of the skull base may extend into the sellar region and involve structures surrounding the sella [3].

Due to the complex neurovascular anatomy of the parasellar region, meningiomas involving the area still represent a formidable task. An understanding of the anatomy of the sella and surrounding structures is essential for surgeons dealing with tumors of the area [4].

The goals of surgery in meningiomas depend on the type of tumor growing. Tumors that grow in globular fashion and displace the surrounding structures can be completely resected with good clinical prognosis. On the contrary, meningiomas that grow "en plaque" frequently show tight encasement of important neurovascular structures, and cannot be completely resected without producing severe neurological deficits. Diffuse invasion of the basal dura, the cavernous sinus, bone, and neuroforamina will make the surgical cure impossible. In these cases, resection of the globular part of the tumor for decompression of neural tissues is the surgical goal.

Depending on the tumor origin, a pattern of displacement of the neurovascular structures can be ex-

pected. Meningiomas arising from the anterior fossa may displace the chiasm posteriorly and the optic nerves laterally, and the A1 segment upward. As the tumor size increases, it may displace the pituitary stalk posteriorly or laterally, invaginate the floor of the third ventricle, and grow into the interpeduncular region [2]. These tumors may be resected by a bifrontal approach, a unifrontal (frontolateral) approach, or a pterional approach [1,2]. We believe that suprasellar meningiomas arising from the planum sphenoidale or tuberculum sellae are best approached by a subfrontal (uni- or bilateral) craniotomy.

Sphenoid wing meningiomas are approached by a frontotemporal (pterional) craniotomy. In case of orbital invasion with proptosis, a wide orbital decompression is indicated. Petroclival meningiomas may invade the parasellar region and the cavernous sinus (Fig. 1) [3]. In our series, 50% of the petroclival meningiomas showed parasellar extension. We prefer a combined subtemporal-presigmoid (transpetrosal) approach for resecting these tumors.

Adequate knowledge about the tumor origin and extension, and use of appropriate skull base approach will increase the completeness of tumor resection and improve postoperative results.

References

1. Al Mefty O, Holoubi A, Rifai A, Fox JL (1985) Microsurgical removal of suprasellar meningiomas. Neurosurgery 16: 364–372
2. Samii M, Ammirati M (1992) Surgery of skull base meningiomas. Springer, Berlin Heidelberg New York, Tokyo
3. Samii M, Tatagiba M (1992) Experience with 36 cases of petroclival meningiomas. Acta Neurochir (Wien) 118: 27–32
4. Tindall GT, Barrow DL (1990) Tumors of the sellar and parasellar area in adults. In: Youmans JR (ed) Neurological surgery, vol 5. Saunders, Philadelphia, pp 3447–3498

Correspondence: M. Samii, M.D., Neurochirurgische Klinik, Krankenhaus Nordstadt, Haltenhoff-strasse 41, D-30167 Hannover, Federal Republic of Germany.

Acta Neurochir (1996) [Suppl] 65: 66–69

Atypical and Malignant Meningiomas: Evaluation of Different Radiological Criteria Based on CT and MRI

R. Verheggen[1], M. Finkenstaedt[2], V. Bockermann[1], and E. Markakis[1]

[1]Clinic of Neurosurgery and [2]Department of Neuroradiology, University of Göttingen, Göttingen, Federal Republic of Germany

Summary

The following are our results of a retrospective analysis of 214 patients, operated on meningiomas, in order to investigate radiological criteria of malignancy. Among these cases there were 31 patients with a histologically confirmed diagnosis of malignant subtypes. As uncertain signs of malignancy of ensuing radiological features are an irregular enhancement of contrast-media and the size of cerebral edema.

Based upon CT and MR images we have developed a standardised, computerised evaluation method which enables us to study in detail the internal architecture of meningiomas.

Keywords: Atypical and malignant meningioma; computerised evaluation method; NMR (MRI); CT (computed tomography).

Introduction

Meningiomas constitute approximately 14 to 19% of the primary brain tumours [4,12]. Although these growths are considered to be chiefly benign, well circumscribed, slow in growth and potentially curable in the majority of cases, atypical and malignant variants occur in 7.2% respectively 2.4% of the meningiomas [8]. A rapid re-growth after apparently thorough removal is not an essential criteria but a blunt or grossly apparent brain invasion is indicative of malignancy. Histopathologically, atypical and malignant meningiomas are characterised by focal necrosis, high mitotic activity with an increased cell proliferation rate, nuclear pleomorphism of different degrees, hypercellularity, loss of architecture and the phenomenon of "sheeting" [7].

As most of the meningiomas are clinically asymptomatic [10], computed tomography (CT) and magnetic resonance imaging (MRI) facilitate the incidental diagnoses more frequently [2]. A preoperative differentiation between benign meningiomas and the malignant counterparts seems to be most desirable.

In our retrospective investigation, we intended to evaluate radiological attributes of malignant subtypes using a standardised computer-controlled procedure.

Patients and Methods

Between January 1987 and April 1994, 214 patients suffering from meningiomas –142 women and 72 men – underwent neurosurgery. Among them there were 18 female and 13 male patients with a histopathologically confirmed diagnosis of atypical meningioma with incipient signs of anaplasia (n=16) and overtly anaplastic or malignant (n=15) meningiomas. The tumor grading corresponds to the proposal outlined by the World Health Organisation (WHO; 2nd meeting, Zurich: 28 March–1 April 1990) [5].

In 11, respectively 12 patients, a primary atypical or primary malignant meningioma was diagnosed. In the remaining 8 cases (5 atypical/3 malignant meningiomas) up to three neurosurgical interventions preceded the final diagnosis of a secondary atypical or malignant meningioma. Except in the case of an atypical meningioma of the cerebellopontine-angle, all of the atypical and malignant variants had a supratentorial localisation in common.

CT and MR-images were analysed by a Kontron® image processing system using a clearly defined evaluation method. After input of the characteristic image sequences by a TV camera and digitalisation all the data were filed. Subsequently, the regions of interest (ROI) of selected CT and MRI scans were enlarged and scaled. Then the sectional line was defined and the respective grey values were depicted. Finally the real length of the sectional line was calculated and the data of the corresponding images were documented.

Results

Radiological Investigation and Representative Case Reports

The diagnosis of a meningioma was provided by CT and/or MRI; optionally, the vascular supply of the tumours was clarified by angiography or angio-MRI. Usually meningiomas are hyperdense on unenhanced CT scans in 74–78% [3,14] and less than 15% appear isodense. In the presence of a cystic component they rarely present themselves hypodense [3]. In nearly 20% a calcification of the tumour is discernible [6]. After application of contrast-media, the benign meningiomas enhance homogeneously [6,11] and are well demarcated. The majority of the meningiomas are isointense on T1 weighted MR-images and the benign subtypes reveal a homogenous enhancement after application of Gadolinium DTPA [3]. On T2 weighted images they impress either as a isointense or hyperintense mass lesion. An heterogeneous and irregular enhancement of contrast media is mostly related to focal or intensive necrosis and indistinct margins of the tumour are suspicious of an atypical or malignant variant in both CT and MRI [9,13].

Case 1

A thirty-eight-year-old woman complained about a mild hypakusis, orthostatic hypotension and several syncopes. As the internal examination was inconspicuous, a CT was arranged revealing a benign falx meningioma (2 × 2.5 × 2 cm) in the region of the left precentral gyrus enhancing homogeneously by contrast-media. The analysis of the grey values demonstrated an increase and levelling of contrast-media enriched areas (Fig. 1).

Case 2

Three weeks prior to hospital admittance, a sixty-one-year-old woman sustained several focal motor seizures in her right leg. She also suffered from nausea and impaired faculty of concentration. A computed tomography exposed a fronto-parietal meningioma of the convexity extending to the falx without an invasion of the sagittal sinus.

The computerised controlled procedures illustrated an heterogeneous distribution of grey values with a concave depression appropriate to the central necrosis and spikes due to tiny dots of calcifications. The application of contrast-media heightened the contrast and accentuated the central necrosis. Almost the same results were attained using T1 weighted MR images after application of contrast-media (Figs. 2–4).

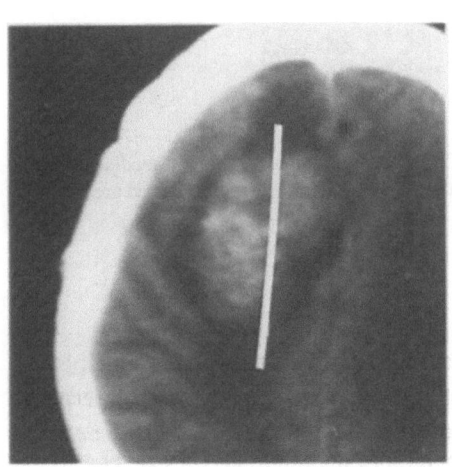

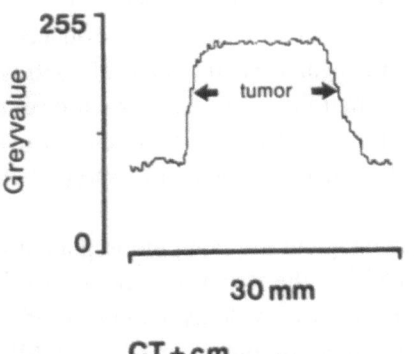

CT + cm.

Fig. 1. Case 1. Contrast enhanced CT scan demonstrating a falx meningioma with the corresponding grey values forming a plateau

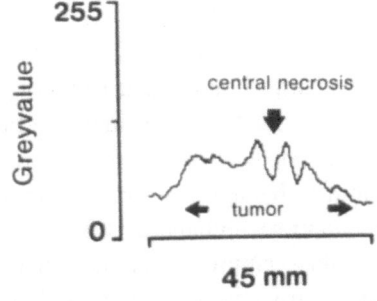

CT native

Fig. 2. Case 2. Native CT of an atypical meningioma with the referring grey values: The central necrosis causes a decline of the curve whereas small dots of calcifications led to spikes

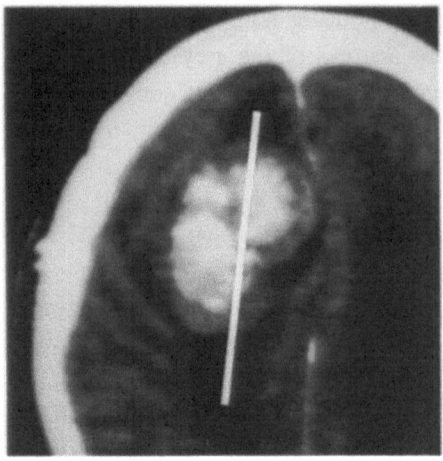

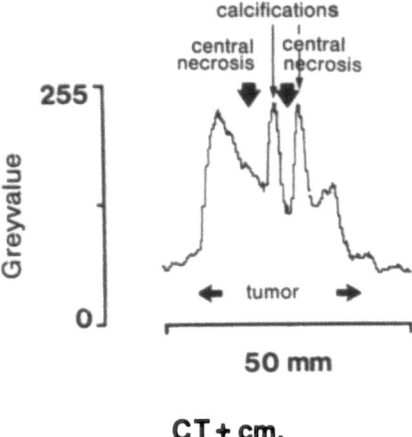

CT + cm.

Fig. 3. Case 2. CT after application of contrast-media: Note the amplification of contrasts in heterogenously enhanced areas

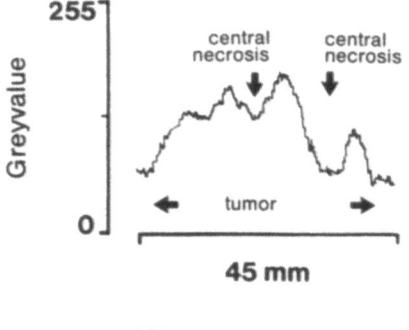

MRI post GD

Fig. 4. Case 2. Gadolinium enhanced, T1 weighted coronal MR images of the above-mentioned patient showing a nearly similar curve of grey values in comparison with Fig. 3

Discussion

Inasmuch as clinical features were not helpful in differentiating benign from malignant counterparts, several attempts were undertaken in order to describe radiological criteria of malignancy [1,9,13]. The size of the tumour related to cerebral edema, or a heterogeneous enhancement due to intratumoral calcifications were indefinite radiological signs.

The presented procedure of a standardised analysis of grey values using enlarged and scaled ROIs of selected CT and MRI scans, facilitates the interpretation and favours a thorough analysis of the internal architecture of meningiomas. Benign subtypes display on both native and contrast enhanced CTs minor differences of grey values within the tumour whereas the malignant counterparts reveal clearly a dissimilar distribution of grey values.

In general, the application of contrast-media is followed by enriched contrasts simultaneously level-ling out contrast-enriched areas with contrasting calcifications. Spreading of grey values are always based on central necrosis or intratumorous calcifications.

Besides the results of Elster and co-workers [1] describing a greater than 75% success rate in discriminating histological subtypes, we cannot ascertain an advantage of MRI over CT in a further differentiation of meningiomas. CT and MRI are of the same value regarding a preoperative distinction between benign and malignant meningiomas based on the evidence of focal necrosis, an irregular enhancement of contrast-media, indistinct margins and the appearance of "mushrooming" [7], which reflect the extension over the cerebral surface from the globoid aspect of the tumour [9].

The major advantage of our computerised analysis consists in objectifying the visual perception of grey values in ROIs. To prove the validity of our method we intend to employ this computerised CT and MRI analysis in a prospective study.

Acknowledgements

The authors thank Dr. G. Latta, Mrs. R. Vania and J. Donohoe for the critical revision of the manuscript and S. Germeyer for preparing the pictorial material.

References

1. Elster AD, Challa VR, Gilbert TH, Richardson DN, Contento JC (1989) Meningiomas: MR and histopathologic features. Radiology 170: 857–862
2. Kallio M, Sankila R, Hakulinen T, Jääskeläinen J (1992) Factors affecting operative and excess long-term mortality in 935 patients with intracranial meningioma. Neurosurgery 31: 2–12
3. Kazner E, Wende S, Grumme T, Stochdorph O, Felix R, Claussen C (1989) Computed tomography and magnetic resonance tomography of intracranial tumors: a clinical perspective. Springer, Berlin Heidelberg New York Tokyo
4. Kepes JJ (1982) Meningiomas. Biology, pathology, and differential diagnosis. In: Masson monographs in diagnostic pathology. Masson, New York, pp 112–123
5. Kleihues P, Burger PC, Scheithauer BW (1993) The new WHO classification of brain tumours. Brain Pathol 3: 255–68
6. Lee SH, Rao K (1987) Cranial computed tomography and MRI. McGraw-Hill, New York
7. Mahmood A, Caccamo DV, Tomecek FJ, Malik GM (1993) Atypical and malignant meningiomas: a clinicopathological review. Neurosurgery 33: 955–963
8. Maier H, Öfner D, Hittmair A, Kitz K, Budka H (1992) Classic, atypical, and anaplastic meningioma: three histopathological subtypes of clinical relevance. J Neurosurg 77: 616–623
9. New PJF, Hesselink JR, O'Carroll CP, Kleinman SM (1982) Malignant meningiomas: CT and histological criteria, including a new CT sign. AJNR 3: 267–276
10. Rausing A, Ybo W, Stenflo J (1970) Intracranial meningioma: a population of ten years. Acta Neurol Scand 46: 102–110
11. Reeder MM, Bradley WG (1993) Gamuts in neuroradiology. Springer, Berlin Heidelberg New York Tokyo
12. Rubinstein LJ (1982) Tumors of the central nervous system. Atlas of tumor pathology, Second Series, Fascicle 6. Armed Forces Institute of Pathology, Washington, DC
13. Vassilouthis J, Ambrose J (1979) CT scanning appearances of intracranial meningiomas. An attempt to predict the histological features. J Neurosurg 50: 320–327
14. Wende S, Aulich A, Kretschmar K, Grumme T, Meese W, Lange S, Steinhoff H, Lanksch W, Kazner E (1977) Die Computertomographie der Hirngeschwülste. Radiologie 17: 149–156

Correspondence: R. Verheggen, M.D., Clinic of Neurosurgery, University of Göttingen, Robert-Koch-Str. 40, D-37075 Göttingen, Federal Republic of Germany.

Acta Neurochir (1996) [Suppl] 65: 70–72
© Springer-Verlag 1996

Prognostic Significance of Nuclear DNA Content in Human Meningiomas: A Prospective Study

J. Meixensberger, M. Janka, A. Zellner, W. Roggendorf, and **K. Roosen**

Neurochirurgische Klinik und Poliklinik, Universität Würzburg, Würzburg, Federal Republic of Germany

Summary

Flowcytometric DNA analyses were performed to study the correlation between alterations of nuclear DNA content and clinical aggressive tumour behaviour in 134 cranial meningiomas. Forty-one meningiomas revealed an aneuploid DNA content with a distribution of n=24 in benign, n=12 in atypical and n=5 in anaplastic tumours. Aneuploid DNA content was correlated with a significantly higher amount of histomorphological criteria like evidence of mitoses, necrosis, infiltration and increased cellularity. There was a significantly higher Ki 67 proliferation index in the aneuploid meningiomas in comparison to the diploid tumour group. The rate of aneuploid cell – lines was increased in recurrent tumours. No tumour recurrence could be found in diploid meningiomas during follow up (mean 37 months, range 22–46 months). However eight of forty-one aneuploid tumours showed meningioma recurrence. Nuclear DNA content has an important signficance in predicting risk of recurrence and poor clinical outcome after benign meningioma surgery.

Keywords: DNA flow cytometry; meningioma; DNA aneuploidy; Ki 67.

Introduction

Histological criteria of usually benign meningiomas are not sufficient to describe the biological course of the tumour. Especially the risk of recurrence shows an unsatisfactory correlation to clinical and pathohistological factors. Although morphological, surgical, radiological and clinical findings correspond to criteria of benignancy, the recurrence rate varies between 9 and 19%. Ironside *et al.* [2] suggested a close relation between nuclear DNA content (measured by flowcytometry) and clinical behaviour in meningiomas studying small series of meningioma patients. Changes of DNA content were also proven to correlate to the clinical course in different tumours (e.g. breast and colorectal carcinomas). The aim of this prospective study was to evaluate quantitative DNA-changes in meningiomas and to correlate this with the clinical course of the patients in order to find a prognostic relevance of flow cytometric results.

Patients and Methods

In a prospective analysis 134 patients with histologically defined cranial meningioma were investigated. All patients were operated consecutively between 1989 and 1991 using microsurgical techniques. To evaluate the quantitative DNA content flow cytometric analysis was performed after a modified protocol of Otto [4]. Native tumour samples were frozen and single cell suspensions were prepared. For the fluorescence analysis of the DNA content of single cell suspensions, cell-nuclei had to be stained with DNA-specific DAPI (4,6-diamino-2-phenylindol). Adenosin-thymidin sequences of the DNA are specific binding-targets for DAPI. Flow cytometry using stained tumour-cell-suspension and defined standard-cell-suspension was combined with a special analysis-software by Rabinovich [5]. This results in DNA-histograms presenting tumour-specific quantitative DNA-changes.

Simultaneously histological subtype and histomorphological criteria were evaluated using the new WHO-classification for meningioma. A qualitative and semiquantitative morphological investigation concentrated on the number of mitoses, infiltrative growth (bone and/or cortex), tightness and pleomorphism of cells as well as tumour necrosis.

To define the biological grade of aggressiveness the proliferation rate was measured immunohistochemically. For this the proliferation marker Ki 67 was used on histopathological paraffin-sections. Ki 67 monoclonal antibody recognises only nuclear antigene expressed in cells within a proliferative cellcycle status.

The clinical course of patients was observed at least annually by neurological investigation and CT-scan. Patients with atypical or malignant meningioma were followed up in 6 months intervals.

Results

Out of the 134 investigated tumours 127 could be included in this study. In 7 meningiomas flow-analysis showed a coefficient of variation (CV) out of the double standard deviation. 108 tumours occurred as primary tumours, 19 meningiomas were operated as tumour-recurrency.

Table 1 summarizes the tumour distribution according to the WHO meningioma classification. Additionally the number of DNA-aneuploidies for each histological subtype is listed. Forty-one meningiomas revealed an aneuploid DNA content with a distribution of n=24 in benign, n=12 in atypical and n=5 in anaplastic tumours. The distribution of hypo- and hyperploid cells was not significantly different.

Figure 1 shows the correlation of histomorphological criteria and aneuploid DNA content. The increased

number of aneuploidies goes along with the significant higher amount of histomorphological criteria like evidence of mitoses, infiltration of cortex and increased cellularity. Infiltration of the cortex was seen in n=33 (primary tumour), increased number of mitoses was observed in n=28 (primary tumour) and n=11 (recurrent tumour). It is well known for these findings to indicate aggressive clinical behaviour (Fig. 3) (p<0.001). An also evident increased number of tumour necrosis was again correlated with a significant higher amount of aneuploidies.

Additionally there was a significant higher Ki 67 proliferation index in the aneuploid meningiomas (1.6% Ki 67 positivity). In comparison only 0.15% of the diploid tumour group showed a Ki 67 positive result (p<0.05) (Fig. 2).

During a clinical follow up mean-period of 37 months (range 22–46 months) no tumour-recurrency was observed in the diploid meningioma group. 8 out of 41 aneuploid tumours showed meningioma recurrence within the described follow up period (Kaplan-Meier p<0.001).

Table 1. *Distribution of Aneuploid Cell-Lines According to Histological Subtypes of Meningiomas*

Histological subtype	Number of tumours	Tumours with aneuploid cell-lines
Benign	105	25 (23.8%)
Atypical	17	12 (67.5%)
Anaplastic	5	5 (100%)

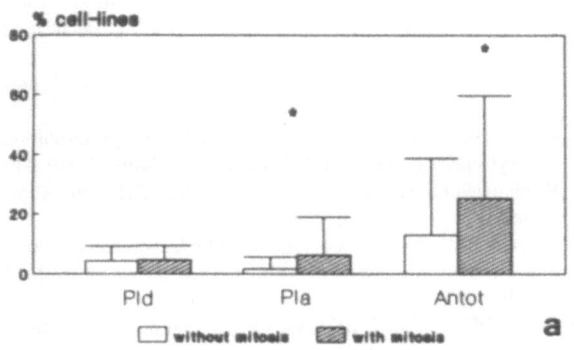
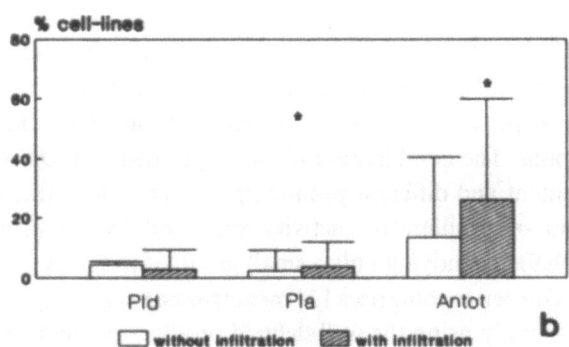

Fig. 1. (a) Histomorphological criteria and aneuploid cell-lines. Proliferations-index PId, PIa and percentage aneuploid cell-lines in primarily operated meningiomas with (n=29) and without (n=81) mitoses. * p<0.05. (b) Histomorphological criteria and aneuploid cell-lines. Proliferations-Index PId, PIa and percentage aneuploid cell-lines in primarily operated meningiomas with (=33) and without (n=77) infiltration.

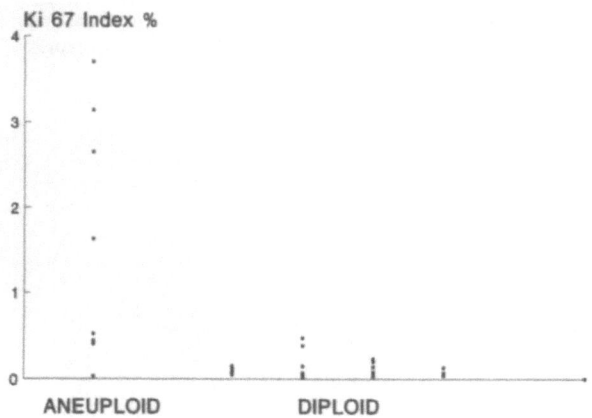
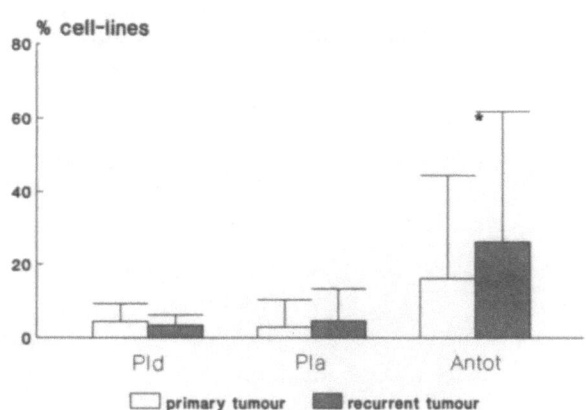

Fig. 2. Ki 67-index and aneuploid cell-lines

Fig. 3. Primary and recurrent meningiomas and aneuploid cell-lines

Discussion

Intracranial meningiomas are in more than 90% of the cases histologically benign tumours. In most of the cases a complete tumor removal can be obtained. Considering this, operations on meningiomas are indicated in patients with intracranial hypertension and major neurological deficits. Despite new facilities in microsurgery allowing complete tumour surgery, there is still an impressively high number of tumour recurrences. A number of groups worked on prediction of recurrence by clinical and surgical aspects during the last 20 years with still unsatisfying results.

Since the first cytogenetic changes in meningiomas has been described by Zang *et al.* [8] a relation between biological tumour behaviour and remarkable genetic findings has been presumed. But the pure karyotyping of meningiomas was not able to offer a prognostic tool in handling meningioma patients.

Taylor *et al.* [7] introduced in 1980 a new quantitative evaluation of DNA content in different stages of the cell cycle using flow cytometry. Laerum *et al.* [3] described first in 1981 clinical applications of this technique and in 1985 Christov *et al.* [1] published results of ploidy abnormalities in brain tumors. Since a positive correlation between aneuploidy and tumour course was found in different carcinomas, flow cytometry had to be proven as a possible prognostic tool in meningioma. The combination of flow cytometry of DNA content and different proliferative markers as indicators of proliferative activity was used by Shapiro (1989) [6] studying only a small group of tumours.

Our series comprises 134 meningiomas in a prospective study using the well-defined proliferative marker Ki 67 and results from DNA flow cytometry. Regarding technical problems, which led to widespread results throughout the literature, we were able to base our data on a new cytometry software [5]. Using CV for control of quality the presented data show best figures (mean CV < 2) when compared to the literature.

In benign meningioma 22.9% show aneuploidy, in atypical meningioma 64.7% and in anaplastic meningioma 100% have aneuploidy in cytometric investigation. A significant number of aneuploidies could be found in tumours with histomorphological criteria indicating proliferation such as nuclear pleomorphism, high rate of mitoses and areas of necrosis. The evaluation of extended necrosis and increased tumour aneuploidy has to be done carefully. Some of the tumours have been treated with embolisation preoperatively.

Also infiltration of the dura or the bone was strongly correlated with aneuploidy. Thus, a significant change in DNA content in meningioma cells compared to standard suspension may indicate an aggressive tumour biology and the need for a careful clinical follow-up of the patients. The also significantly higher rate of aneuploidies in recurrent meningiomas underlines this conclusion.

References

1. Christov K, Zapryanov Z (1985) Flow cytometry in brain tumors. I. Ploidy abnormalities. Neoplasma 33: 49–55
2. Ironside JW, Battersby RDE, Dangerfield VJM *et al.* (1986) DNA in meningioma tissues and explant cell cultures: a flow cytometric study with clinicopathological correlates. J Neurosurg 66: 588–594
3. Laerum OD, Farsund T (1981) Clinical application of flow cytometry: a review. Cytometry 2: 1–13
4. Otto F (1990) DAPI-staining of fixed cells for high-resolution flow cytometry of nuclear DNA. Methods Cell Biol 33: 105–110
5. Rabinovich PS: Handbook Multicycle Vers. 2.5 (Phoenix Flow Systems)
6. Shapiro HM (1989) Flow cytometry of DNA content and other indicators of proliferative activity. Arch Pathol Lab Med 113: 591–597
7. Taylor IW, Milthorpe BK (1980) An evaluation of DNA fluorchromes, staining techniques and analysis for flow cytometry. J Histochem Cytochem 28: 1224–1232
8. Zang KD, Singer H (1967) Chromosomal constitution of meningiomas. Nature 216: 659–663

Correspondence: PD Dr. J. Meixensberger, Neurochirurgische Klinik und Poliklinik, Universität Würzburg, Josef-Schneider-Strasse 11, D-979080 Würzburg, Federal Republic of Germany.

Acta Neurochir (1996) [Suppl] 65: 73–76
© Springer-Verlag 1996

Correlation Between Cytogenetic and Clinical Findings in 215 Human Meningiomas

W.I. Steudel[1], **R. Feld**[1], **W. Henn**[2], and **K.D. Zang**[2]

[1]Neurochirurgische Klinik der Universität des Saarlandes and [2]Institut für Humangenetik, Universitätskliniken des Saarlandes, Homburg, Federal Republic of Germany

Summary

The management of meningiomas remains a major challenge to the neurosurgeon because patients having this common benign tumor can be cured effectively by surgical resection. But there are a number of meningiomas that have an aggressive course and tend to recur. Predicting the recurrence of meningiomas has often been mentioned in the context of histology or surgical techniques and some approaches considered. However, the recurrence rate remains between 10% and 20%, even after total removal. To improve the care of patients with problematic meningiomas, 215 different human meningiomas were collected between 1976 and 1993 and cytogenetically analyzed using standard techniques.

140 patients could be observed for 1 to 17 years after complete tumor removal, whereby 21 tumors (15%) displayed one or more recurrences during that long-term observation period. The tumors were classified according to different clonal abnormalities: we observed recurrence in 10 (9.1%) of 111 (79.2%) tumors having a normal karyotype or typical monosomy of chromosome 22, whereas 9 (69.7%) of 13 tumors with pronounced hypodiploidy and 3 (35%) of 8 tumors with a hyperdiploid karyotype recurred. A loss of the short arm of chromosome 1 was identified in 6 meningiomas with a recurrence rate of 60%.

Our observations show that the correlation between meningiomas and recurrence is highly significant (p=0.002) and that these tumors require special treatment in addition to surgical skill.

Keywords: Meningioma; recurrence; cytogenetic studies.

Introduction

The recurrence in meningiomas has been widely discussed in the literature [1,3,5,7,9]. Although generally considered benign the behavior of these tumors is unpredictable and characterized by frequent recurrences. Although most of these tumor recurrences can be attributed to incomplete primary excision, recurrence rates after complete tumor resection range from 3 to 21% [5,9]. Certain histological subtypes and microscopic features have been described but little correlation between these characteristics and the growth rate could be established [1] making the prediction of recurrence by clinical or histopathological means inadequate.

Studies on the correlation between karyotypic findings and histopathological or clinical features are few [2,8,10] but severe chromosomal abnormalities reported in meningiomas raise questions regarding the role of these abnormalities in the biological behavior of the tumor. In order to determine whether cytogenetic findings can be correlated with clinical findings such as recurrence, location or histological type more than 200 different meningiomas were cytogenetically analysed between 1976 and 1993.

Material and Methods

This is a retrospective study of 215 patients with meningiomas who were diagnosed and operated at the Neurochirurgische Universitätsklinik in Homburg/Saar. Since 1976 tumor tissue was obtained during open surgery and placed in cell culture. Chromosome preparation and GTG banding of meningiomas were performed according to standard protocols. The constitutional karyotype of the patients was determined in epidermal cell cultures and was normal in all cases.

Patients were incorporated in the follow-up part of this study with the following criteria:

1) histological confirmation of meningioma,
2) surgically complete tumor resection,
3) postoperative follow-up of at least 1 year.

Complete surgical excision was defined as Simpson Grades I and II [9] which means at least macroscopically complete removal of the tumor with diathermic coagulation of its dural attachment. To confirm the intraoperative sight of excision a postoperative computed tomography showing complete tumor removal was necessary to enter the follow-up. Data were obtained from hospital notes and radiographic studies as well as by contacting patients.

75 patients lost to follow-up or with incomplete tumor removal were excluded from the follow-up part of this study, The follow-up ranges from 1 year to 17 years. Mean follow-up was 7.55 years (91 $ 34 months) for the group with recurrent and 6.67 years (80 $ 29 months) for the non-recurrent meningiomas. Overall, the mean time of tumor recurrence was 4.12 years (49 $ 38 months).

Histological diagnosis of benign versus atypical and malignant meningiomas was done according to the WHO classification [11]. Mantel-Haenszel Chi-square and Fisher's exact Test (2-Tail) were used to correlate clinical and cytogenetical findings.

Results

A total of 215 meningiomas was cytogenetically analysed. Tumor karyotype, sex, location and patho-histology are shown in Tables 1 and 2.

Due to cytogenetical features 5 groups could be performed:

Group 0 with a normal karyotype (46, XX/XY);
Group 1 with monosomy of chromosome 22 (45, XX/ XY, –22);
Group 2 with a hypodiploid karyotype with/without loss of chromosome 22;
Group 3 with a hyperdiploid karyotype with/without loss of chromosome 22;
Group 4 with structural aberrations with/without loss of chromosome 22.

In 106 cases (49.4%) a normal karyotype was found. Monosomy 22 without association with other abnor-

Table 1. *Correlation Between Clinical and Cytogenetic Findings.* Age, sex, histology (n=215)

Feature	Group 0 46, XX/XY no. %	Group 1 45, XX/XY-22 no. %	Group 2 Hypodiploid no. %	Group 3 Hyperdiploid no. %	Group 4 Pseudodiploid no. %	Total no. %
Cases	106 (49,4)	59 (27,4)	17 (7,9)	11 (5,1)	22 (10,2)	215 (100)
Male	27 (24,5)	17 (28,8)	6 (35,3)	1 (9,1)	1 (4,5)	52 (24,2)
Female	79 (75,5)	42 (71,2)	11 (64,7)	10 (90,9)	21 (95,5)	163 (75,8)
Age m	54,0 ± 10,6	56,7 ± 8,4	55,7 ± 7,9	0	63	
Age f	51,2 ± 12	58,7 ± 14,3	59,7 ± 6,7	54,5 ± 10,2	54,3 ± 9,2	
Histology (WHO)						
Common type	100	55	12	9	21	197 (91,6)
Atypical	4	2	4	2	1	13 (6,0)
Anaplastic	2	2	1			5 (2,4)

Table 2. *Localization (n=215)*

Feature	Group 0 46, XX/XY no. %	Group 1 45, XX/XY-22 no. %	Group 2 Hypodiploid no. %	Group 3 Hyperdiploid no. %	Group 4 Pseudodiploid no. %	Total no. %
Localization						
Convexity	30 (28,3)	21 (35,5)	6 (35,2)	6 (5,4)	9 (41,0)	72 (33,4)
Parasagittal	17 (16,0)	11 (18,6)	3 (17,6)	3 (27,3)	1 (4,5)	35 (16,3)
Tub. sellae	9 (8,5)	1 (1,7)				10 (4,7)
Olfactory groove	16 (15,1)	3 (5,1)				19 (8,8)
Med. sphenoid	6 (5,7)	1 (1,7)	1 (5,9)	1 (9,1)	2 (9,1)	11 (5,1)
Lat. sphenoid	10 (9,4)	2 (3,4)	2 (11,8)		4 (18,2)	18 (8,4)
Tentorial	5 (4,7)	4 (6,8)	2 (11,8)		1 (4,5)	12 (5,6)
Posterior fossa	6 (5,7)	8 (13,6)			2 (9,1)	16 (7,4)
Intraventricular			1 (5,9)			1 (0,5)
Multiple	5 (4,7)	3 (5,1)	2 (11,8)	1 (9,1)	1 (4,5)	12 (5,6)
Spinal	2 (1,9)	5 (8,5)			2 (9,1)	9 (4,2)

Table 3. *Cytogenetic and Clinical Results.* Follow up group (n=140)

Group	Karyotype	Patients	Aged < 45	WHO II/III	Recurrence
0	normal	71 (50%)	19 (27%)	3 (4%)	7 (10%)
1	monosomy 22	40 (29%)	5 (13%)	4 (10%)	3 (8%)
2	hypodiploid	10 (7%)	2 (20%)	5 (50%)	7 (70%)
3	hyperdiploid	8 (6%)	2 (25%)	3 (38%)	5 (63%)
4	pseudodiploid	11 (8%)	2 (18%)	1 (9%)	0 (0%)
	total	140 (100%)	30 (21%)	16 (11%)	22 (16%)

malities was a frequent event in 59 cases (27.4%). The loss of the short arm of a chromosome 1 (del 1p) was observed in 6 cases and was listed under group 2. As shown in Table 1, neither the clinical nor the histological features of the meningiomas varied significantly between the cytogenetic types.

A follow-up of more than 1 year could be performed in 140 cases. Table 3 shows the distribution according to different groups of karyotype and some clinical features. Recurrence was observed in 21 cases with an overall recurrence rate of 15% after total excision. Meningiomas with isolated monosomy 22 (Group 1) had a rate of recurrence of 8% and no meningioma with pseudodiploid karyotype (Group 4) recurred during the time of observation. There was a highly significant correlation between recurrence rate and hypo- or hyperdiploid karyotypes (Groups 2 and 3) with a more than 5-fold elevated recurrence rate (50%). Nevertheless there was no difference if an additional loss of a chromosome 22 was involved or not. Partial monosomy 1p was encountered in 6 meningiomas. Two of them were hypodiploid (both recurred) and two of them were hyperdiploid (both recurred as well) whereas two had balanced translocations (Group 4) and did not recur.

The recurrence rate within 5 years from surgical excision was 1 of 7 (14%) for group 0 meningioma (this one was an anaplastic meningioma), 3 of 7 (43%) within 10 years and 3 of 7 (43%) late recurrences after 10 years whereas Groups 2 and 3 meningiomas had early recurrences of 8 of 12 (66.7) within the first 5 years, 3 of 12 (25%) within 10 years and late recurrence was observed in only one case (8.3%).

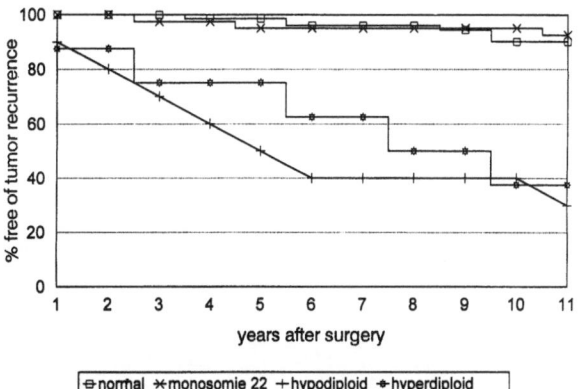

Fig. 1. The cumulative proportion of all the patients free of recurrence against the number of years since the first surgery

Discussion

Since the early days of neurosurgery meningiomas have been characterized by frequent recurrences. While the extent of tumor excision is agreed to be the most important prognostic factor in meningiomas, the prognostic significance of the clinical or histological character of the growth remains controversial. Many attempts have been made to classify histological subtypes and introduce microscopic features such as focal necrosis or increased mitotic rates [4,6,9] as associated with clinically aggressive behavior, other authors found little correlation between histological characteristics and recurrence [1]. In our series there was a tendency for atypical or malignant meningiomas to correlate with recurrence although this was not statistically significant. We further found no correlation between recurrence and clinical findings such as sex, age and the tumor location (Table 2) although it is generally believed that tumor side is correlated with recurrence rate [9]. As other authors [5] have pointed out correctly, these studies often predate the era of microscopical surgery and the standard use of computed tomography making it difficult to separate real recurrence from regrowth due to subtotal tumor resection.

While analysing the karyotype in meningiomas we have been able to distinguish between 5 groups with similar chromosomal abnormalities and found that hypo- or hyperdiploid meningiomas displayed an up to 10 times elevated recurrence rate. The mean time to tumor recurrence in these cytogenetically suspicious meningiomas was within the first 5 years from surgery, which displays the aggressive biological behavior of these tumors and is in contrast to the meningiomas with normal karyotypes or monosomy 22 which whenever recur far later.

Monosomy 22 has been confirmed by several authors [2,8,10] to be a frequent abnormality in meningioma but these meningiomas had the lowest rate of recurrence with 3 of 40 (7.5%) as well displayed signs of malignancy in only 4 of 55 (7.3%) cases making them the "typically" benign meningioma. This observation is exceeded by the higher recurrence rate of the meningiomas with normal caryotypes which the authors believe may be explained by the artifact of in vitro overgrowth of normal cells in some cultures of actually atypical meningiomas. Abnormal karyotypes different from monosomy 22 may indicate or initiate aggressive tumor characteristics similar to what has been observed in human malignant tumors and it seems that cytogenetic analysis of meningioma helps to identify "difficult" meningiomas before morphological changes lead to the diagnosis of an atypical or ananplastic meningioma.

The results of this study suggest that the analysis of the karyotype in meningioma may be used as an indicator of prognostic value in identifying a subgroup of patients at higher risk of recurrence who may be considered for special follow-up and treatment.

References

1. Adegbite AB, Khan MI, Paine KWE, Tan LK (1983) The recurrence of intracranial meningiomas after surgical treatment. J Neurosurg 58: 51–56
2. Casalone R, Simi P, Granata P, Minelli E, Giudici A, Butti G, Solero CL (1990) Correlation between cytogenetic and histopathological findings in 65 human meningiomas. Cancer Genet Cytogenet 45: 237–243
3. Chan RC, Thompson GB (1984) Morbidity, mortality, and quality of life following surgery for intacranial meningiomas. A retrospective study in 257 cases. J Neurosurg 60: 52–60
4. Jellinger K, Slowik (1975) Histological subtypes and prognostic problems in meningioma. J Neurol 208: 279–298
5. Mahmood A, Qureshi NH, Malik GM (1994) Intracranial meningiomas: Analysis of recurrence after surgical treatment. Acta Neurochir (Wien) 126: 53–58
6. Mahmood A, Caccamo DV, Tomecek FJ, Malik GM (1993) Atypical and malignant meningiomas: a clinicopathological review. Neurosurgery 33: 955–963
7. Mirimanoff RO, Dosoretz DE, Linggood RM, Ojemann RG, Martuza RL (1985) Meningioma: analysis of recurrence and progression following neurosurgical resection. J Neurosurg 62: 18–24
8. Saadi AA, Latimer F, Madercic M, Robbins T (1987) Cytogenetic studies of human brain tumors and their clinical signi-ficance. II. Meningioma. Cancer Genet Cytogenet 26: 127–141
9. Simpson D (1957) The recurrence of intracranial meningiomas after surgical treatment. J Neurol Neurosurg Psychiatry 20: 22–39
10. Zang KD (1982) Cytological and cytogenetical studies on human meningioma. Cancer Genet Cytogenet 6: 249–274
11. Zülch KJ (1979) Histological typing of tumors of the central nervous system. Geneva, World Health Organisation

Correspondence: Dr. W.I. Steudel, Neurochirurgische Universitätklinik, D-66421 Homburg/Saar, Federal Republic of Germany.

Acta Neurochir (1996) [Suppl] 65: 77–81

Surgical Results of Spinal Meningiomas

J. Klekamp and **M. Samii**

Medical School of Hannover, Neurosurgical Clinic, Nordstadt Hospital, Hannover, Federal Republic of Germany

Summary

We report on 94 spinal meningiomas in 88 patients operated between September 1977 and August 1994 which were followed for up to 13 years (mean 24 ± 35 months). Complete tumour resection led to postoperative improvement of every preoperative deficit or symptom. En plaque, recurrent, anterior, and low thoracic or lumbar meningiomas were likely to be resected incompletely. Partial tumour removal, arachnoid scarring, primary dural suture, recurrent meningiomas, and male sex were independent factors predisposing to clinical recurrence. Cauterization instead of resection of the tumor matrix was not associated with a higher recurrence rate.

Keywords: Meningioma; spinal cord neoplasm; arachnoiditis.

Introduction

Spinal meningiomas carry a favourable prognosis if resected completely [1,2,4–7,10–12,16,18,19,21,23,25, 28,29]. However, a number of factors such as arachnoid changes or the growth pattern of the meningioma have not been taken into account in the majority of papers. We will present in our analysis which factors determine the resectability of spinal meningiomas as well as the neurological outcome and predispose for tumour recurrence.

Material and Methods

During the period from 1977 to 1994 a total of 94 spinal meningiomas in 88 patients were operated in the Department of Neurosurgery at the Nordstadt Hospital in Hannover, Germany. Case records and neuroradiological findings were evaluated. According to intraoperative findings, encapsulated and en plaque meningiomas were distinguished. En plaque meningiomas were characterized by absence of a tumour capsule, violation of tissue planes and extensive dural infiltration.

A recurrence was defined clinically as neurological deterioration after surgical treatment – independent of evidence for tumour growth on MRI-scan. Surgical morbidity was defined as a new permanent neurological deficit or an aggravation of a pre-existing symptom without subsequent recovery.

Means are presented plus/minus the standard deviation. For statistical analyses Student's t-tests for paired or unpaired variables were used provided the Komolgorov–Smirnov test indicated normal data distribution. The rate of recurrence was determined by Kaplan–Meier analysis [14]. Statistical differences for recurrence rates were determined using the log rank test. A multiple regression analysis was performed to determine which factors were of significant importance for the resectability of a particular tumour, the probability of a recurrence, and postoperative outcome. A difference was considered significant if a p-value of 0.05 was reached.

Results

The average patient age was 57 ± 16 years (range 17 to 84 years). Female sex predominated by a factor of 2.6:1 (65 females and 23 males). The average history until admission for surgery was 23 ± 33 months. The majority of patients demonstrated a slowly progressive course. One patient presented a history of 18 years. For the majority of patients, the first symptoms noted were pain and/or dysesthesias (52%). The remaining patients complained about motor weakness and gait disturbances (37%) or sensory changes (10%) as the first symptoms or sign.

On admission, the clinical picture was different with problems of motor power and/or gait in 76% of patients. 24 patients presented with acute worsening of neurological function rendering them unable to walk. The average Karnofsky score [15] on admission was 61 ± 15 (range 30 to 90), indicating significant clinical symptoms for the majority of patients. With en plaque growing meningiomas, 56% complained mainly about pain and/or dysesthesias and only 45% described gait disturbances or motor weakness as their major concern. For meningiomas with a well defined capsule, the corresponding figures were 17% and 83%, respectively.

78% of tumours demonstrated a tumour capsule, whereas 18% showed an en plaque growth pattern

Table 1. *Localization of Spinal Tumors*

Spinal level	Tumour matrix		
	Dorsal	Lateral	Anterior
Cervical			
– with capsule	5	10	12
– en plaque	–	2	2
– total	5	12	14
Thoracic			
– with capsule	15	25	5
– en plaque	2	10	3
– total	17	35	8
Lumbar			
– with capsule	–	–	1
– en plaque	1	–	–
– total	1	–	1
All			
– with capsule	20	35	18
– en plaque	3	12	5
– total	23	47	23

en plaque en plaque growth pattern.

Table 2. *Complications*

Type of complication	With capsule	En plaque
Infection	1	–
Aseptic meningitis	1	–
CSF-leak, pseudomeningocele	1	2
Haemorrhage	1	–
Instability	1	–
Aspiration pneumonia	1	–
Myocardial infarction	1	–
Postoperative permanent deficit	–	1
	7 (9%)	3 (18%)

(Table 1). Three patients had multiple meningiomas and 8 patients had neurofibromatosis type 2. 13% of operations dealt with a recurrent tumour. We observed a significant difference for cervical tumours to be situated anteriorly more often than thoracic meningiomas (52% and 13%, respectively; p=0.0006) (Table 1). On the other hand, thoracic meningiomas were more commonly found laterally than cervical tumours (57% and 28%, respectively; p=0.0006) (Table 1). Eight tumours (9%) showed extradural extension. 22% presented evidence of arachnoid scarring (11% of previously unoperated tumours and all recurrent tumours; p<0.0001). Encapsulated tumours showed a significantly lower tendency for arachnoid scarring than en plaque growing tumours (14% and 55%, respectively; p=0.0003). Complete tumour removal was achieved for 87% of previously unoperated patients and 13% of recurrent tumours. A duraplasty was inserted in 47% of cases. For patients with complete removal, the tumour matrix was resected in 69% and cauterized in 31%. Incomplete tumour removal was more likely with an en plaque growth pattern, a recurrent (p<0.01), a low thoracic or lumbar, and an anterior meningioma (p<0.05, multiple regression analysis).

Complications occurred in 10% of patients (Table 2). Five patients died during the period of follow-up. Two patients with meningiomas of the upper cervical spine died within 30 days of surgery due to aspiration pneumonia or myocardial infarction (surgical mortality 2.2%). The remaining 3 patients died within 1 year of surgery after complete tumour re-

moval without evidence of a recurrence due to cardiac or respiratory diseases.

Patients were followed for a mean period of 24±35 months (maximum 13 years). In general, patients benefited from surgery considerably. Of 24 patients unable to walk preoperatively, 19 regained this ability within a few days or after rehabilitation programs within a year.

For encapsulated tumours, significant improvements were seen for all symptoms and signs and the Karnofsky score (t-test for paired variables, p<0.05). For en plaque growing tumours, improvements were seen for motor weakness and gait ataxia only.

A preliminary worsening of preoperative neurological function was observed for 11% of patients. They regained their preoperative neurological level after a maximum interval of 6 months. A permanent postoperative neurological deficit was observed for one patient with a recurrent en plaque growing meningioma (surgical morbidity 1.1%).

A high Karnofsky score 1 year after surgery was observed after complete tumour removal, no arachnoid scarring, first rather than recurrent surgery, a high preoperative Karnofsky score (p<0.01, respectively), and no tumour recurrence (p<0.05, multiple regression analysis). The length of history or the growth pattern of the tumour did not show an independent, significant influence on the postoperative Karnofsky score.

In total, 10 patients demonstrated a clinical recurrence. According to a Kaplan–Meier analysis this corresponds to a recurrence rate of 23.5% within 13 years. Of these, the majority were due to recurrent tumour growth or growth of residual tumour. However, 2 patients showed a progressive neurological deterioration in the absence of tumour growth. In these patients, significant arachnoid scarring was considered responsible as demonstrated with MRI in one case and verified by surgery in the other.

Independent risk factors for postoperative neurological deterioration (= clinical recurrence) were partial tumour removal, preoperative arachnoid scarring (p<0.01, respectively), primary dura suture instead of a duraplasty, surgery on a recurrent tumour, and male sex (p<0.05, respectively; multiple regression analysis).

After complete resection, patients remained free of a clinical recurrence in 84% of cases whereas all patients with partially removed meningiomas demonstrated a clinical recurrence within 5 years of surgery (log rank test: p=0.03 at 12 months). For patients with complete tumour removal, no significant difference was seen whether the tumour matrix had been resected or cauterized (83% and 89%, respectively). For meningiomas with a well defined capsule, Kaplan–Meier analysis showed that 88% remained free of a clinical recurrence. The corresponding figure for en plaque meningiomas was 20% (log rank test: p=0.01 at 10 months postoperatively). If arachnoid scarring was not encountered at surgery, 77% of patients remained free of a clinical recurrence compared to 47%, if arachnoid scarring had been present (log rank test: p=0.0006 at 6 months). The first operation on a spinal meningioma carried a risk of clinical recurrence in 23% of patients compared to 71% for recurrent meningiomas (log rank test: p=0.009 at discharge).

Discussion

In terms of age and sex distributions as well as localization of the meningiomas similar findings were reported previously [1–7,9–12,16,18,19,21,22,25–29, 32]. In general, surgery leads to a favourable outcome [1,2,4–7,9–12,16,18,19,21,23,25,28,29]. However, even with modern imaging techniques such as MRI, we still observe a considerable delay in diagnosis of this tumour. A clinical history of sensory changes, dysesthesias and pain accompanied or followed by gait ataxia should alert the physician to the possibility of a spinal tumour. In these patients, a thorough neurological examination should lead to radiological studies and establish the diagnosis [19,21,28].

As is the case with other spinal tumours, nothing is gained by postponing surgery once an extramedullary tumour is diagnosed. The great majority follow a gradually progressive course of neurological deterioration [1,2,5,7,14,16,18,19,23,28]. A significant number of patients demonstrate a rapid decline in neurological function prior to admission [1,5,16,23,28]. Interestingly, this holds especially for meningiomas which present a well developed capsule. The reason for rapid deterioration in these patients may be interference with spinal cord blood flow due to compression of the cord.

Meningiomas without a capsule tend to infiltrate dura and/or arachnoid. In this manner, they may become symptomatic by pain and dysesthesias due to meningeal irritation rather than by signs of cord compression, i.e. gait disturbances and motor weakness.

Clinical presentation, intraoperative findings and postoperative outcome call for a separate analysis of en plaque growing meningiomas. They tended to be associated with a significant amount of arachnoid scarring rendering surgery more difficult and hazardous [8,30]. It may be extremely difficult to distinguish between dura, arachnoid, tumour, and spinal cord – especially if the meningioma had been operated on before, as was the case in 35% of them. In such instances, we rather left some tumour behind than risking a severe aggravation of neurological deficits.

The higher proportion of incomplete resections with low thoracic or lumbar meningiomas can be explained in part by the higher rate of en plaque growing meningiomas in this localization. With meningiomas in the conus area or anterior meningiomas, preservation of the arterial blood supply of the cord prompted surgeons to take special care in microsurgical dissection. Again, it was decided rather to leave tumour remnants behind than to risk neurological deficits [6,7,12,29].

About every fifth tumour was found to be associated with significant arachnoid scarring. These arachnoid changes were not restricted to recurrent meningiomas after a previous operation. 11% of previously unoperated tumours presented with arachnoid scarring as well, in particular en plaque growing meningiomas [30]. Arachnoid scarring was found to be associated with a worse prognosis as far as neurological outcome and risk for a recurrence was concerned [30]. A similar trend was seen for intramedullary tumours and dysraphic malformations [17,27]. Apart from postsurgical arachnoid scarring, we consider mechanical irritation of the leptomeninges due to spinal cord movements at the tumour site as the most likely explanation for arachnoid scarring [20]. Arachnoid scarring may contribute to neurological symptoms independently from the meningioma. It may lead to pain, dysesthesias or progressive spinal cord damage due to interference with spinal cord blood flow [13,20,26] and meningeal irritation. It may even mimic a tumour recurrence with progressive motor weakness and sensory disturbances [20,30,31] as seen in two patients of this study.

Now that the great majority of patients can be operated for spinal tumours with low morbidity we consider it to be an important task to minimize the problem of postoperative arachnoid scarring which may lead to very troublesome dysesthesias for a significant number of patients or even progressive neurological deterioration. Whenever arachnoid scars are found during surgery, we recommend careful sharp dissection of the arachnoid avoiding bipolar coagulation as much as possible and decompression of the spinal canal with a fascia lata graft after the tumour has been removed. In our experience, this technique limits the amount of postoperative arachnoid scarring considerably compared to primary dura closure without a graft, possibly due to the irrigative effect of CSF flow, which is maintained with an enlarged subarachnoid space [17,25,27].

Complications occurred in 10% of patients, with a surgical mortality of 2.2% and permanent surgical morbidity in 1.1%. One patient with a high cervical meningioma acquired fatal aspiration pneumonia postoperatively, one patient died of myocardial infarction within 30 days of the operation. Another 3 patients died during the period of follow-up due to cardiac or respiratory diseases. In the literature, mortality rates vary between 1% and 5.3% [9,10,12,18, 19,23,28].

Clinical symptoms and signs responded favourably to surgery even if severe neurological deficits had developed preoperatively [1,2,4–7,10,12,16,18,23,29], especially after complete tumour removal, no arachnoid scarring, and a good preoperative neurological condition [4,5,12,16,18].

Again, encapsulated tumours tended to do better postoperatively than en plaque meningiomas. According to our multiple regression analysis, this is mainly due to the fact that a higher proportion of en plaque growing meningiomas had been removed incompletely and arachnoid changes were present more often [30].

Solero et al. [28] reported a recurrence rate of 6% for completely and 17% for partially removed tumours. McCormick et al. [21] described a recurrence rate of 10–15%. However, these figures underestimate the number of recurrences because a Kaplan–Meier analysis was not performed so that the varying follow-up times were not accounted for. We have defined a recurrence on clinical and not on radiological criteria and used a Kaplan–Meier analysis to determine recurrence rates. In total, 11% of patients in this series demonstrated a clinical recurrence. This figure corresponds to a recurrence rate of 23.5% in 13 years as determined by Kaplan–Meier analysis. Using Kaplan–Meier

analysis but radiological criteria to determine a recurrence, Mirimanoff [22] gave a lower figure of 0% after 5 years but 13% after 10 years.

Multiple regression analysis demonstrated that partial tumour removal, arachnoid scarring, primary dural suture, and surgery on a recurrent meningioma were significant risk factors for a clinical recurrence [22]. Of clinical relevance is the finding that cauterization as well as resection of the tumour matrix led to identical results in terms of recurrence rates [1,19,28]. Disturbing is the fact that only 20% of en plaque meningiomas remained in a stable neurological condition postoperatively [30]. These tumours remain to be a major challenge to the neurosurgeon in any localization due to absence of a tumour capsule and ill defined if not absent arachnoidal planes. Other than with encapsulated tumours, extended dura resections are probably required to achieve a higher rate of patients free of recurrence. Likewise, the number of recurrences is very high for recurrent meningiomas. The first surgeon to operate on a meningioma will have the best chance to provide a long-term benefit or even cure for the patient. With each subsequent operation, the chances for stabilizing or even improving the patient's condition will diminish as dissection planes are more difficult to define each time an additional operation is required.

References

1. Boccardo M, Ruelle A, Mariotti E (1985) Personal experience with the surgery of spinal meningiomas. Ital J Neurol Sci 6: 29–35
2. Bret P, Lecuire J, Lapras C, Deruty R, Dechaume JP, Arsaad A (1976) Les meningiomas intra-rachidiens. Reflexions apropos d'une serie de 60 observations. Neurochirurgie 22: 5–22
3. Bull JWD (1953) Spinal meningiomas and neurofibromas. Acta Radiol 40: 283–300
4. Chern SH, Lin SM, Tseng SH, Tu YK, Yang LS, Kao MC, Hung CC (1993) Prognostic factors of intraspinal neurilemmoma and meningioma with severe preoperative motor defictis. J Formos Med Assoc 92: 227–230
5. Ciapetta P, Domenicucci M, Raco A (1988) Spinal meningiomas: prognosis and recovery factors in 22 cases with severe motor deficits. Acta Neurol Scand 77: 27–30
6. Cushing H, Eisenhardt L (1938) Meningiomas. Their classification, regional behaviour, life history, and surgical end results. Thomas, Springfield
7. Davis RA, Washburn PL (1970) Spinal cord meningioma. Surg Gynecol Obstet 131: 15–21
8. Freidberg SR (1972) Removal of an ossified ventral thoracic meningioma. Case report. J Neurosurg 37: 728–730
9. Goldhahn WE, Schmidt U (1989) Das spinale Meningiom. Zentralbl Neurochir 50: 18–23
10. Gräwe A, Siedschlag WD, Nisch G, Schulz MR (1986) Meningiome des Spinalkanals. Klinik und Langzeitergebnisse. Zentralbl Neurochir 47: 139–143

11. Horrax G, Poppen JL, Wu WR, Weadon PR (1949) Meningiomas and neurofibromas of the spinal cord. Certain clinical features and end results. Surg Clin N Am 29: 659–665

12. Iraci G, Peserico L, Salar G (1971) Intraspinal neurinomas and meningiomas. A clinical survey of 172 cases. Int Surg 56: 289–303

13. Kang JK, Kim MC, Kim DS, Song JV (1987) Effects of tethering on regional and spinal cord blood flow and sensory evoked potentials in growing cats. Childs Nerv Syst 3: 35–39

14. Kaplan EL, Meier P (1958) Nonparametric estimation from incomplete observations. J Am Sat Assoc 53: 457–481

15. Karnofsky DA, Burchenal JH (1949) The clinical evaluation of chemotherapeutic agents in cancer. In: MacLeod CM (ed) Evaluation of chemtherapeutic agents. Columbia University Press, New York, pp 191–205

16. Katz K, Reichental E, Israeli J (1981) Surgical treatment of spinal meningiomas. Neurochirurgia 24: 21–22

17. Klekamp J, Raimondi AJ, Samii M (1994) Occult dysraphism in adulthood: clinical course and management. Childs Nerv Syst 10: 312–320

18. Kunicki A, Maciejak A (1965) Results of operative treatment in the 154 cases of extramedullary meningiomas and neurinomas. Acta Medica Pol 6: 397–404

19. Levy WJ Jr., Bay J, Dohn D (1982) Spinal cord meningioma. J Neurosurg 57: 804–812

20. Mackay R (1939) Chronic adhesive spinal arachnoiditis. A clinical and pathologic study JAMA 112: 802–808

21. McCormick PC, Post KD, Stein BM (1990) Intradural extramedullary tumors in adults. In: Stein BM, McCormick PC (eds) Neurosurgery clinics of North America, Vol 1, no 3. Intradural spinal surgery. Saunders, Philadelphia, pp 591–608

22. Mirimanoff RO, Dosoretz DE, Lingood RM, Ojemann RG, Martuza RL (1985) Meningioma: analysis of recurrence and progression following neurosurgical resection. J Neurosurg 62: 18–24

23. Namer IJ, Pamir MN, Benli K, Saglam S, Erbengi A (1987) Spinal meningiomas. Neurochirurgia 30: 11–15

24. Nittner K (1976) Spinal meningiomas, neurinomas and neurofibromas. In: Vinken PJ, Bruyn GW (eds) Handbook of clinical neurology, Vol 20. Tumours of the spine and spinal cord, Part II. North Holland, Amsterdam, pp 177–322

25. Onofrio BM (1978) Intradural extramedullary spinal cord tumors. Clin Neurosurg 25: 540–555

26. Rogers L (1955) Tumors involving the spinal cord and its nerve roots. Ann R Coll Surg Engl 16: 1–29

27. Samii M, Klekamp J (1994) Surgical results of 100 intramedullary tumors in relation to accompanying syringomyelia. Neurosurgery 35: 865–873

28. Solero CL, Fornari M, Giombini S, Lasio G, Oliveri G, Cimino C, Pluchino F (1989) Spinal meningioma: review of 174 operated cases. Neurosurgery 25: 153–160

29. Souweidane MN, Benjamin V (1994) Spinal cord meningiomas. Neurosurg Clin N Am 5: 283–291

30. Stechison MT, Tasker RR, Wortzman G (1987) Spinal meningioma en plaque. Report of two cases. J Neurosurg 67: 452–455

31. Vloeberghs M, Herregodts P, Stadnik T, Goossens A, D'Haens J (1992) Spinal arachnoiditis mimicking a spinal cord tumor: a case report and review of the literature. Surg Neurol 37: 211–215

32. Wenker H, Reuter F (1986) Spinal tumours: a multicenter study of the Deutsche Gesellschaft für Neurochirurgie. Adv Neurosurg 14: 81–95

Correspondence: Jörg Klekamp, M.D., Medical School of Hannover, Neurosurgical Clinic, Nordstadt Hospital, Haltenhoffstr. 41, D-30167 Hannover, Federal Republic of Germany.

Acta Neurochir (1996) [Suppl] 65: 82–85

Microsurgical Management of Ventral and Ventrolateral Foramen Magnum Meningiomas

H. Bertalanffy[1], **J.M. Gilsbach**[1], **L. Mayfrank**[1], **H.M. Klein**[1], **T. Kawase**[3], and **W. Seeger**[2]

Departments of Neurosurgery, [1]Technical University of Aachen and [2]University of Freiburg, Federal Republic of Germany, and [3]Keio University, Tokyo, Japan

Summary

The authors report their experiences gained from 19 patients with ventral or ventrolateral foramen magnum meningiomas operated on via the dorsolateral, suboccipital transcondylar access route. It is emphasized that the microsurgical management of these lesions includes two important aspects which increase the safety of the procedure: a meticulous preoperative planning based on the microanatomical details of each patient, as well as an individualized tailoring of the surgical approach. There were no deaths, and, in the past 5 years, no neurological complications in this series. Gross total removal of the tumour was achieved in each case. It is concluded that microsurgical removal of ventral or ventrolateral foramen magnum meningiomas with this technique constitutes a safe and recommendable procedure.

Keywords: Foramen magnum; meningioma; microsurgery; surgical anatomy; transcondylar approach.

Introduction

Meningiomas located at the ventral or ventrolateral rim of the foramen magnum have a complex relationship to the surrounding neurovascular structures, i.e. to the lower brain stem and/or the upper cervical cord, to the lower cranial and/or upper cervical nerves, to the vertebro-basilar system and sometimes to the jugular bulb. This problematic location is one of the main difficulties neurosurgeons operating on them have to deal with. The present communication discusses the most important aspects of the microsurgical management of these lesions which includes the preoperative planning as well as the surgical technique.

Microsurgical Anatomy

The surgeon who operates upon a ventral foramen magnum meningioma must be thoroughly familiar not only with the microsurgical anatomy of the foramen magnum region [3,5,8,10,15], but, above all, with the individual morphological features of each patient undergoing surgery. MRI provides details about the lesion's size, site, extent and relationships to the surrounding neurovascular structures. A thin slice CT scan taken with bony algorithms provides information about the individual configuration of the skull base. This knowledge is complemented by a three-dimensional reconstruction of spiral CT images [13], and shows, when performed with bolus injection of contrast medium, the relationships between tumour, skull base and vertebral artery (Fig. 1). Additionally, a superselective arteriogram demonstrates the vascular supply which is usually provided by extradural branches of the carotid or vertebral arteries.

Since the anatomy of the dorsolateral craniospinal junction is very complex, orientation during the extradural stage of surgery is facilitated by paying attention to at least 5 anatomical landmarks [5–7]. *1. The dural entrance of the vertebral artery* which must be dissected free. *2. The posterior condylar emissary vein* which is located in the posterior condylar emissary canal and drains into the jugular bulb. This vein must be coagulated and divided to avoid bleeding from the jugular bulb. *3. The medial rim of the distal sigmoid sinus* which must be exposed over 2 or 3 mm, indicates the lateral limit of the surgical access route (Fig. 1). *4. The hypoglossal canal* which courses in anterolateral direction through the occipital condyle. *5. The jugular tubercle* which is a bony structure of variable size, surrounded by the cranial nerves 9, 10 and 11 medially and superiorly, by the jugular bulb laterally and by the

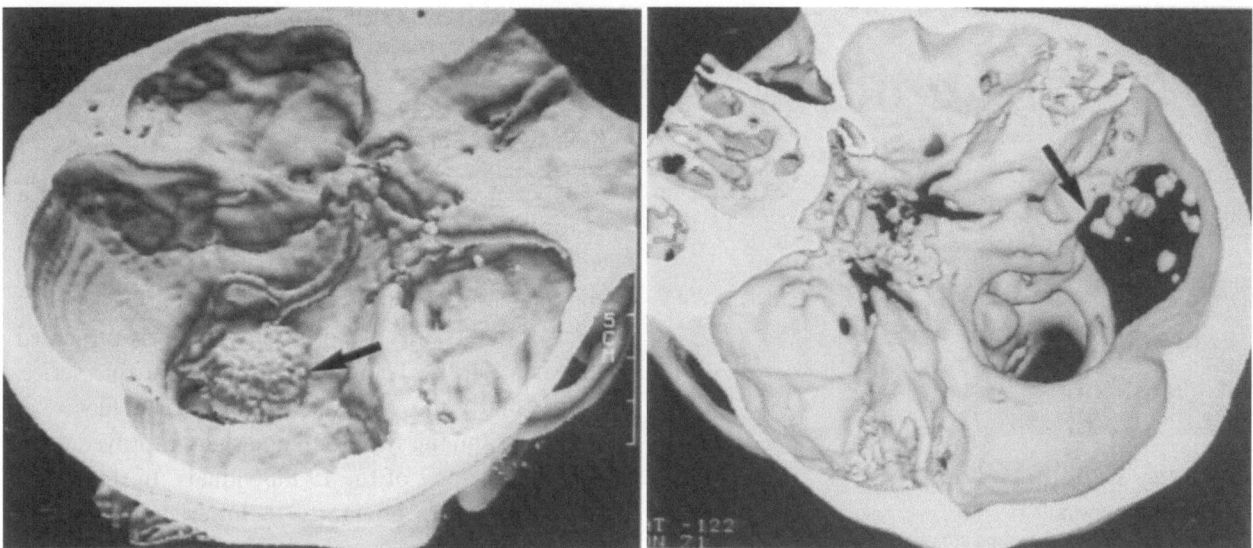

Fig. 1. Left: Preoperative three-dimensional reconstruction of the spiral CT of a 74-year-old female showing a meningioma attached to the right ventrolateral rim of the foramen magnum (arrow). The right vertebral artery is caudally displaced by the tumour. Right: Postoperative image of the same patient performed with the same technique. The craniotomy extends laterally to the medial rim of the sigmoid sinus (arrow). The mastoid process and the C1 lamina have remained intact

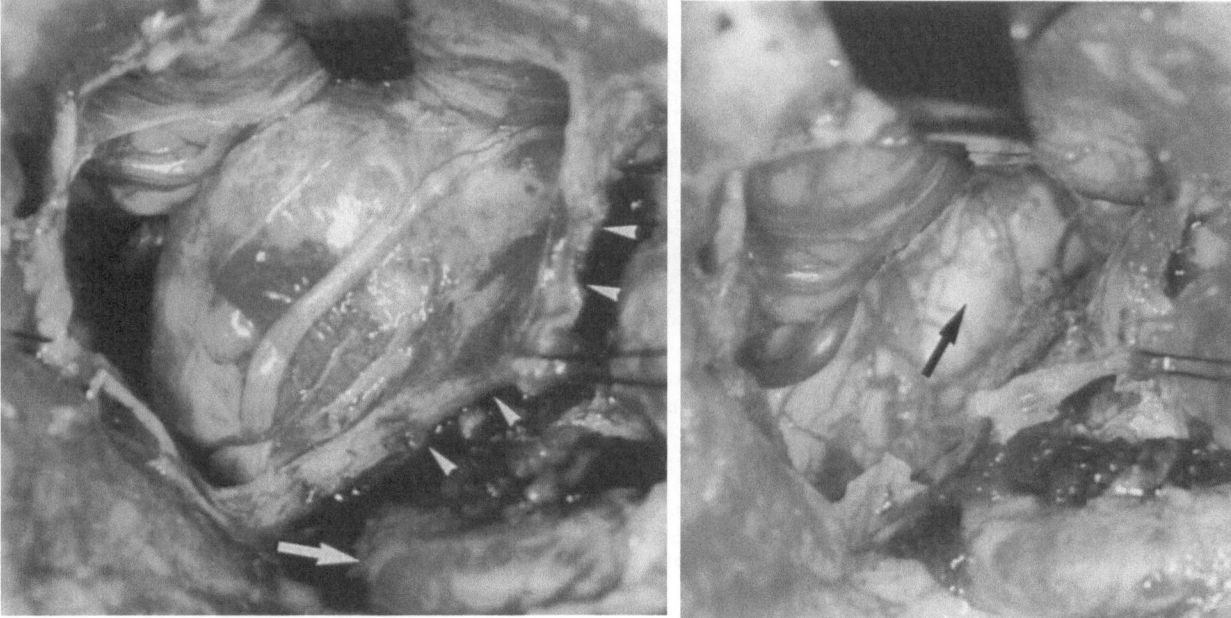

Fig. 2. Left: Intraoperative view of the meningioma shown in Fig. 1. After removal of the medial portion of the lateral atlantal mass and occipital condyle (arrowheads), opening of the dura and slight elevation of the cerebellum exposes the tumour and its ventrolateral attachment to the dura. The arrow points to the dural entrance of the vertebral artery. Right: Following total removal of the meningioma, the distorted brain stem becomes visible (arrow)

intracanalicular portion of the hypoglossal nerve inferiorly and posteriorly.

Surgical Technique

The procedure, performed either in the sitting or in the lateral part bench position, is composed of the following surgical steps: *1.*

Muscular division and partial exposure of the posterolateral atlantal arch. In our opinion, the longitudinal skin incision between midline and mastoid process [4] is superior to the U-shaped incision [18–20] because it saves time and produces less muscular trauma. *2. Exposure of the posterior condylar fossa and the horizontal portion of the vertebral artery.* Medially, the exposure includes the dorsolateral rim of the foramen magnum, laterally it extends to the medial rim of the mastoid process. The vertebral artery is readily identified within the sulcus posterior to the lateral atlantal mass. (Fig. 2). *3. Partial*

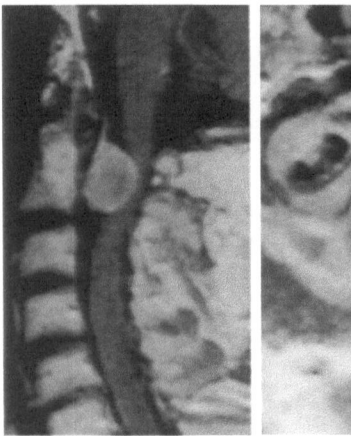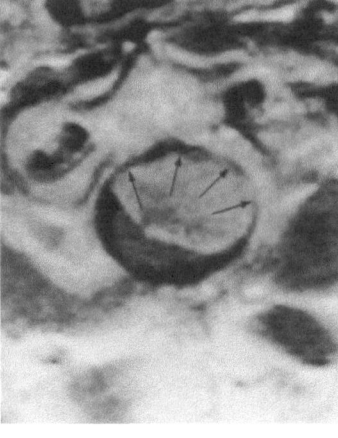

Fig. 3. Left: Preoperative MRI (midsagittal section) of a 72-year-old female demonstrating a ventral meningioma at the C1-C2 level. Right: Axial view of the same tumour showing its ventral and left ventrolateral dural attachment (arrows). The spinal cord is displaced posteriorly and contralaterally

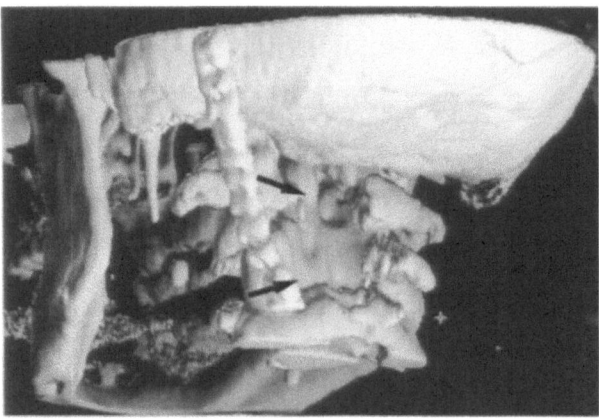

Fig. 4. Postoperative spiral CT scan (three-dimensional reconstruction) of the patient shown in Fig. 3. The tumour has been approached by a left-sided C1 and C2 hemilaminectomy (arrows)

suboccipital craniectomy. In our experience, it is sufficient to enlarge the foramen magnum by 2 or 3 cm posterolaterally and to slightly expose the medial rim of the distal sigmoid sinus (Fig. 1). C1 and C2 hemilaminectomy are performed only in the cases with cervical location of the meningioma (Figs. 3 and 4). *4. Partial resection of the occipital condyle and unroofing of the hypoglossal canal.* This surgical step is necessary to expose the dura anterolaterally to the brain stem, which is the most frequent site of tumor attachment and vascular supply. *5. Resection of the jugular tubercle.* This is the final, however, most difficult extradural surgical step. A sufficient amount of the jugular tubercle must be resected in order to obtain a straight-line view anterior to the lower brain stem. *6. Dural opening and intradural stage.* After obtaining complete extradural haemostasis, the dura is opened close to the medial rim of the sigmoid sinus in the upper portion, and close to the medial rim of the dural entrance of the vertebral artery in the lower. The tumour, and particularly its dural attachment, are thus exposed (Fig. 2), allowing for devascularization of the tumour by successively coagulating the region of attachment. Piecemeal removal of a readily accessible, devascularized meningioma within a dry surgical field can then be accomplished with a high degree of safety, despite the close relationships to

the lower brain stem, the rootlets of the lower cranial nerves as well as the vertebral artery and its intradural branches (Fig. 2).

Patients and Surgical Results

Between 1986 and 1994, 19 patients harbouring a ventral or ventrolateral foramen magnum meningioma were treated microsurgically. There were 14 females and 5 males with a mean age of 59 years. The cases of some of these patients have been previously presented elsewhere [4,11]. There was no death in this series. Gross total removal of the tumour was achieved in each case, and the symptoms and signs improved postoperatively in 17 of the 19 individuals. In the past 5 years there were no injury to lower cranial nerves or vital vascular elements, no CSF leak, and no craniospinal instabilities.

Discussion

Meningiomas are the most frequent tumours of the foramen magnum [10]. Since modern neuroradiological imaging and meticulous microanatomical studies became available, the neurosurgical management of foramen magnum meningiomas has considerably improved [1,14,17]. A number of surgical approaches has been designed, aimed at exploring the previously inaccessible tumours located anterior to the lower brain stem [2–5,9,11,12,16,18–20]. The rationale of all these approaches is to create a direct access route to the anterior rim of the foramen magnum by drilling away functionally less important portions of the skull base. Considering that the exposure of tumours located ventral to the lower brain stem requires a certain amount of surgical traumatization, we believe that one of the main goals of surgery upon ventral or ventrolateral foramen magnum meningiomas should be the attempt of minimizing the surgical traumatization and, consequently, the number of potential sources for complications. To apply a standardized approach in each patient would mean to undertake in many cases unnecessary surgical steps. Instead, every effort should be made to obtain maximum of effectiveness with a minimum of invasiveness. In particular, the following surgical steps should therefore be avoided, whenever possible: exposure of the mastoid process and mastoidectomy; exposure of the lateral half of the C2 lamina; complete exposure of the transverse process of the atlas; opening of the transverse foramen of the atlas, division of the C1 root and medial transposition of the vertebral artery; extensive resection of the medial portion of the lateral atlantal mass; extensive resection of the occipital condyle including its articular facet; complete exposure of the distal sigmoid sinus and jugular bulb, and ligation of the sigmoid sinus.

The present results demonstrate that ventral or ventrolateral meningiomas of the foramen magnum can be removed microsurgically via the dorsolateral trans-condylar route without additional neurological impairment and without producing a craniospinal instability. A meticulous preoperative planning according to the microanatomical details of each patient, and particularly an individualized tailoring of the procedure are indispensible for the success of surgery.

References

1. Al-Mefty O (1989) Surgery of the cranial base. Kluwer, Boston, pp 239–258
2. Babu RP, Sekhar LN, Wright DC (1994) Extreme lateral transcondylar approach: technical improvements and lessons learned. J Neurosurg 81: 49–59
3. Baldwin HZ, Miller CG, van Loveren HR, Keller JT, Daspit CP, Spetzler RF (1994) The far lateral/combined supra- and infratentorial approach. J Neurosurg 81: 60–68
4. Bertalanffy H, Seeger W (1991) The dorsolateral, suboccipital, transcondylar approach to the lower clivus and anterior portion of the craniocervical junction. Neurosurgery 29: 815–821
5. Bertalanffy H, Kawase T, Seeger W, Toya S (1992) Microsurgical anatomy of the transcondylar approach to the lower clivus and anterior craniocervical junction. In: Surgical anatomy for microsurgery V. Sci Med Publications, Tokyo, pp 167–175
6. Bertalanffy H, Gilsbach J, Seeger W, Toya S (1994) Surgical anatomy and clinical application of the transcondylar approach to the lower clivus. In: Samii M (ed) Skull base surgery. First Int Skull Base Congr, Hannover 1992, Karger, Basel, pp 1045–1048
7. Bertalanffy H, Gilsbach JM, Mayfrank L, Kawase T, Shiobara R, Toya S (1995) Planning and surgical strategies for early management of vertebral artery and vertebrobasilar junction aneurysms. Acta Neurochir (Wien) 134: 60–65
8. De Oliveira E, Rhoton AL Jr, Peace D (1985) Microsurgical anatomy of the region of the foramen magnum. Surg Neurol 24: 293–352
9. George B, Dematons C, Cophignon J (1988) Lateral approach to the anterior portion of the foramen magnum. Surg Neurol 29: 484–490
10. George B, Lot G, Velut S, Gelbert F, Mournier KL (1993) Pathologie tumorale du foramen magnum. Neurochirurgie 39 [Suppl 1]
11. Gilsbach JM, Eggert HR, Seeger W (1987) The dorsolateral approach in ventrolateral craniospinal lesions. In: Voth D, Glees P (eds) Diseases in the craniocervical junction. de Gruyter, Berlin, pp 359–364
12. Hakuba A, Tsujimoto T (1993) Transcondyle approach for foramen magnum meningiomas. In: Sekhar LN, Janecka IP (eds) Surgery of cranial base tumors. Raven, New York, pp 671–678
13. Klein HM, Bertalanffy H, Mayfrank L, Thron A, Günther RW, Gilsbach JM (1994) Three-dimensional spiral CT for neurosurgical planning. Neuroradiology 36: 435–439
14. Koos WT, Spetzler RF, Pendl G, Perneczky A, Lang J (1985) Color atlas of microneurosurgery. Thieme, Stuttgart, pp 125–128
15. Lang J (1981) Klinische Anatomie des Kopfes. Springer, Berlin Heidelberg New York
16. Perneczky A (1986) The posterolateral approach to the foramen magnum. In: Samii M (ed) Surgery in and around the brain stem and the third ventricle. Springer, Berlin Heidelberg New York Tokyo, pp 460–466
17. Samii M, Knosp E (1992) Approaches to the clivus. Springer, Berlin Heidelberg New York Tokyo, pp 117–163
18. Sen CN, Sekhar LN (1990) An extreme lateral approach to intradural lesions of the cervical spine and foramen magnum. Neurosurgery 27: 197–204
19. Sen CN, Sekhar LN (1991) Surgical management of anterior placed lesions at the craniocervical junction – an alternative approach. Acta Neurochir (Wien) 108: 70–77
20. Spetzler RF, Grahm TW (1990) The far-lateral approach to the inferior clivus and the upper cervical region: technical note. Barrow Neurol Ins Q 6(4): 35–38
21. Vallée B, Besson G, Houidi K, Person H, Dam Hieu PH, Rodriguez V, Périot Ph, Sénécail B (1993) L'extension latérale juxta- ou transcondylienne de la voie sousoccipitale postérieure. Étude anatomique, intérêt chirurgical. Neurochirurgie 39: 348–359

Correspondence: PD Dr. med. Helmut Bertalanffy, Neurochirurgische Klinik, RWTH Aachen, Pauwelsstrasse 30, D-52057 Federal Republic of Germany.

Acta Neurochir (1996) [Suppl] 65: 86–91
© Springer-Verlag 1996

Meningiomas of the Cerebellopontine Angle

C. Matthies, G. Carvalho, M. Tatagiba, M. Lima, and **M. Samii**

Department of Neurosurgery, Nordstadt Hospital, Hannover, Federal Republic of Germany

Summary

Meningiomas of the cerebellopontine angle (CPA) represent a clinically and surgically interesting entity. The opportunity of complete surgical excision and the incidence of impairment of nerval structures largely depend on the tumour biology that either leads to displacement of surrounding structures by an expansive type of growth or to an enveloping of nerval and vascular structures by an en plaque type of growth. As the origin and the direction of growth are very variable, the exact tumour extension in relation to the nerval structures and the tumour origin can be identified sometimes only at the time of surgery. Out of a series of 230 meningiomas of the posterior skull base operated between 1978 and 1993, data of 134 meningiomas involving the cerebellopontine angle are presented. There were 20% male and 80% female patients, age at the time of surgery ranging from 18 to 76 years, on the average 51 years. The clinical presentation was characterized by a predominant disturbance of the cranial nerves V (19%), VII (11%), VIII (67%) and the caudal cranial nerves (6%) and signs of ataxia (28%). 80% of the meningiomas were larger than 30 mm in diameter; 53% led to evident brainstem compression or dislocation and 85% extended anteriorly to the internal auditory canal. Using the lateral suboccipital approach in the majority of cases and a combined presigmoidal or combined suboccipital and subtemporal approaches in either sequence in 5%, complete tumour removal (Simpson I and II) was accomplished in 95% and subtotal tumour removal in 5%. Histologically the meningiotheliomatous type was most common (49%) followed by the mixed type (19%), fibroblastic (16%), psammomatous (7%), hemangioblastic (7%) and anaplastic (2%) types. Major postoperative complications were CSF leakage (8%) requiring surgical revision in 2% and hemorrhage (3%) requiring revision in 2%. While the majority of neurological disturbances showed signs of recovery, facial nerve paresis or paralysis was encountered in 17%, and facial nerve reconstruction was necessary in 7%. Hearing was preserved in 82% with improvement of hearing in 6%.

The variability of tumour extension, the implications and limitations for complete surgical excision are discussed along with the experiences from the literature.

Keywords: Cerebellopontine angle; meningiomas; suboccipital approach; subtemporal approach.

Introduction

Meningiomas are presumed to account for 15% of brain tumours [15]. While tumours of the cerebellopontine angle constitute a special challenge with regard to diagnostic as well as to therapeutic measures, meningiomas of this region impose additional and variable difficulties: At first this is due to their variable origin and matrix, second to their variable growth directions and third because of their unusual biological behavior and special relationship with nerval and vascular structures. Concerning the latter aspect two types of meningiomas have to be differentiated, the expansive type that displaces surrounding structures and the "en plaque" type that grows along neighboring structures and envelopes them [16]. All these aspects are impossible to know before surgery though they are decisive for the difficulty of surgery, for the clinical prognosis and they might be important for the ideal surgical approach and for the indication and the timing of surgery. As a consequence, the patient with a CPA meningioma has a difficult fate and needs specific care with regard to explanation of his situation and future, to means of diagnostic, of control and of treatment.

134 patients with CPA meningiomas were evaluated for their clinical presentation, operative findings and results.

Patients and Methods

Among 230 meningiomas of the posterior fossa related to the cerebellopontine angle (CPA) operated between 1978 and 1993, there were 134 original CPA meningiomas. Evaluation included clinical symptoms, neurological signs, histology, intraoperative findings, early postoperative results and complications. The patients were operated via the suboccipital or the combined suboccipital/sub-temporal approach in the semi-sitting position. Neurophysiological control included brainstem auditory evoked potentials and facial electromyography, in recent cases also median nerve somatosensory evoked potentials (M-SEP) during positioning of the patient and during tumour removal.

Results

Patient Distribution

Among 134 patients with CPA meningiomas there were 27 male (20%) and 107 female (80%) patients. Mean age in the male group was 48.9 years compared to 41.9 years in the female group.

Surgery

In 95% the suboccipital retromastoid approach and in 5% a combined suboccipital/subtemporal approach were used. 4% of the patients had been operated previously. Complete tumour removal including removal or coagulation of the tumor matrix (Simpson I or II) was achieved in 95% and in 5% tumor removal was subtotal. One of those patients, aged 73 with a 5 cm CPA meningioma received partial decompressive surgery, and, after clinical recovery, was reoperated with complete tumour removal three years later; another patient needed reoperation after one year that was subtotal again because of a meningioma en plaque. Another patient suffered of a malignant type of neurofibromatosis with rapid regrowth of skull base neurinomas besides the meningioma, persistent brainstem dysfunction and respiratory disturbance leading to death 4 months postoperatively.

Surgical Management of the Cranial Nerves

In the majority of cases the cranial nerves were preserved anatomically and functionally. The facial nerve was preserved in continuity in 93%; the facial nerve was reconstructed in 3% by intracranial transplantation, and reanimation by hypoglossal-facial anastomosis was carried out in 4% (17). The trigeminal nerve was preserved free of tumor in 94%, preserved with some tumour infiltration in 2%; in 4% with preoperative anaesthesia or severe dysaesthesias it was resected. The caudal cranial nerves were preserved free of tumor in 97 to 98%, they were preserved at an infiltrated state in 2%; the accessory nerve was reconstructed in 1%.

Growth Extension

Anterior extension reached up to the internal auditory canal in 15% and was anterior to the auditory canal in 85%. 33% had grown anteriorly to the cranial nerve bundle (Fig. 1b). At the CPA and at the jugular foramen two different situations were found: either a tumor mass dorsal to the cranial nerves displacing them anteriorly or a tumour mass rostrally to the cranial nerves and extending and dislocating them dorsally (Figs. 1 and 2). Medial extension was without brainstem compression in 47% and in 53% with brainstem

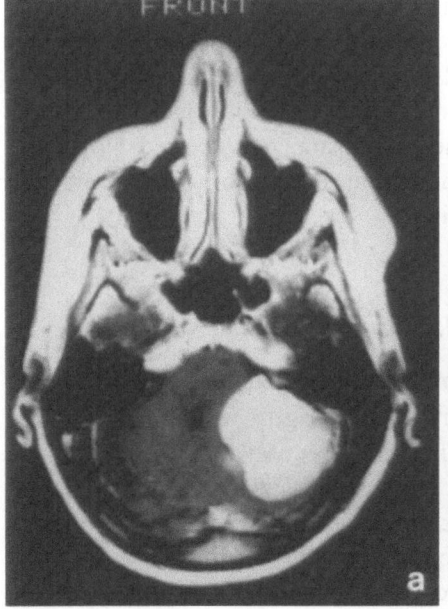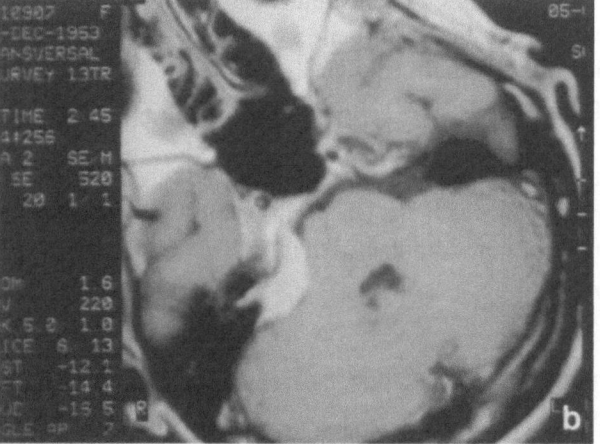

Fig. 1. Two different types of growth extension are demonstrated by Gadolinium-enhanced MRI: (a) a large CPA meningioma with considerable brainstem compression extends to the sigmoid sinus; (b) a small CPA meningioma with tumour growth anterior to the auditory canal seems to involve the auditory canal; only at surgery the direction of nerves dislocation will become evident

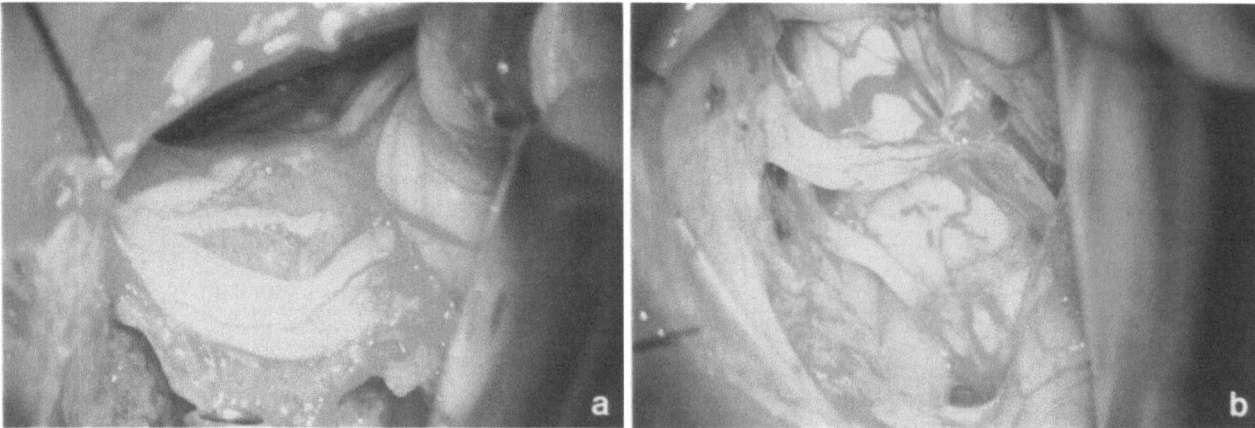

Fig. 2. This left-sided meningioma (view from suboccipital approach in the semi-sitting position) originating anterior to the auditory canal has dislocated the cranial nerves dorsally; while their union at the auditory porus is visible, at their course from the brainstem the cranial nerves are spread apart, the facial and the intermedius nerve are pushed upwards and the eighth nerve backwards (a). Between VII and VIII stepwise enucleation is performed (a) until the matrix from anterior to inferior and posterior of the auditory canal can be coagulated and drilled away and the nerves (b) are preserved free of tumour

compression (Fig. 1a). Lateral extension did not involve the auditory canal in 61%, involved the auditory canal in 26% (Fig. 1b) and led to petrous bone hyperostosis in 13%. Posterior extension reached the auditory canal in 92% and led to sigmoid sinus infiltration in 8%. Superior extension reached up to the auditory canal in 45%, up to the tentorium in 40% and infiltrated the tentorium in 15%. Inferior extension reached to the auditory canal in 55%, the jugular foramen in 29%, the foramen magnum in 16%. Therefore, in 45% involvement of the caudal cranial nerves was possible with anterior or posterior displacement (as in the case of the Vth to VIIIth nerves).

Symptoms and Complaints

72% patients suffered subjective hypacusis for 3.9 years, 58% tinnitus for 3 years, 61% vestibular disturbances for 2.1 years. Trigeminal disturbances occurred in 13%, 7% suffered trigeminal neuralgia for over 3 years and 6% trigeminal hypaesthesia for 1.7 years. 8% had some visual disturbances for 1.2 years; 6% noticed swallowing disturbances for 1 year and 3% noticed some facial asymmetry or weakness. Unspecific symptoms such as headaches were complained of in 36% for a duration of 3 years.

Preoperative Neurological Presentation

The nerves of the extra-ocular muscles were rarely affected; there were 1% paresis of the oculomotor nerve, 2% trochlear paresis, 3% abducens nerve

paresis. The trigeminal nerve was impaired in 19%, the facial nerve in 11% and the cochlear nerve in 67%. 6% had some glossopharyngeal and 3% hypoglossal pareses. Gait disturbances with unsteadiness on walking, pathological Romberg's and Unterberger's tests were found in 28%. Paresis of the upper limb was found in 3% and of the lower limb in 4%.

Postoperative Neurological Presentation

Impairment of function was noted for the trochlear nerve (7%) and abducens nerve (7%). Incidence of facial nerve paresis increased to 17%. Trigeminal nerve function improved within 2 weeks postoperatively and only 12% had some trigeminal disturbance left. Gait function improved, too. Only in 15% gait disturbances were left in the early postoperative period. Upper and lower limb paresis improved, too, with 2% left with reduced function.

Follow-up of Reconstructed Cranial Nerves

The accessory nerve transplant led to reinnervation within 9 months after surgery with continuous improvement of muscle power until now (Grade M3). All the patients with hypoglossal-facial reanimation showed reinnervation leading to complete eye closure and some control of the mouth angle (House–Brackmann Scale III). The patients with facial nerve transplants showed reinnervation with House Grades III to IV [17].

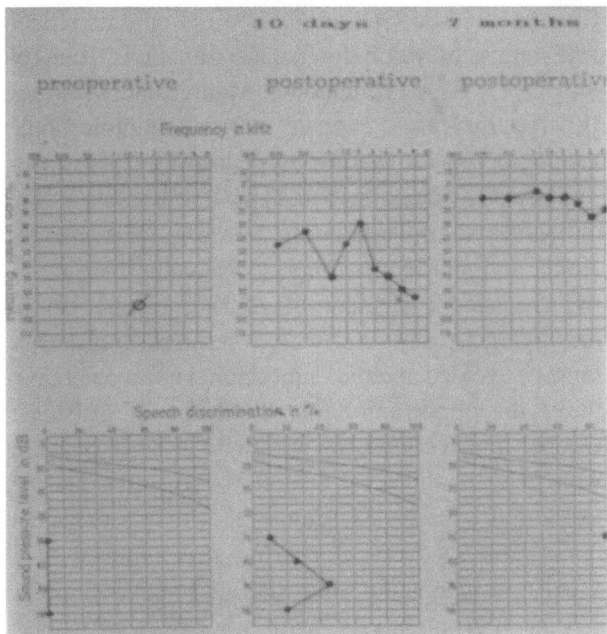

Fig. 3. This 64-year-old patient with right CPA meningioma had no clinically usable hearing preoperatively; early postoperatively, some bad hearing (60dB) and 45% SDS was present that improved further to 10dB and 100% SDS

Cochlear Nerve Function

While preoperatively 14% of patients were deaf, postoperative deafness was present in 25%. Normal to good postoperative hearing was present in 37%, fair hearing in 33% and bad hearing in 5%. Although rarely, there were a few cases (6%) of hearing improvement with recovery from very bad preoperative auditory states (Fig. 3).

Complications

Postoperative hydrocephalus in 3 cases needed temporary CSF-drainage in 1 case and permanent shunt implantation in 2 cases. Paradoxical CSF fistula occurred in 7%; in 3% compressive dressing and in 4% compressive dressing and additional lumbar drainage solved the problem. CSF fistula needed surgical revision in 0.7%, and subcutaneous CSF collection needed revision in 1.5%. Acute hemorrhage occurred in 3% and subacute hemorrhage in 1%, revision was indicated in 2%. Pneumonia occurred in 3%. There were no cases of meningitis.

Discussion

Incidence

Among our patients operated between 1978 and 1993, within 765 intracranial meningiomas 134, i.e. 18%, were located in the CPA and diagnosed after an average of 4 years of symptoms. Granick *et al.* [7] found an incidence of CPA meningiomas of 10% among all intracranial meningiomas and a long duration until tumour diagnosis. Mirimanoff *et al.* [13] found an incidence of 14% of CPA location among 225 intracranial meningiomas. With regard to proportion to acoustic neurinomas, the most common lesions of the CPA, the relation was 134 versus 1000 neurinomas (13%); the reports in the literature are similar to this [18].

In a comparative analysis of acoustic neurinomas and meningiomas Catz *et al.* [3] found that neurinomas present mainly (94%) with the complaint of hearing loss, while meningiomas often (60%) present with nonspecific pains in the head or neck. In our patients the incidence of headaches (36%) was higher than in acoustic neurinoma patients (12% within 1000 cases).

The rate of trigeminal nerve involvement is especially high in meningiomas, 7% in this study, and higher than in neurinomas (3.3% with 1000 cases). Bullitt *et al.* [2] reported on 16 patients with trigeminal pain due to intracranial tumours within 2000 (0.8%); according to his experience atypical trigeminal pain is rather due to middle fossa processes whereas CPA tumors cause rather classical trigeminal neuralgia. Cheng *et al.* [4] found in an analysis of almost 3000 patients with facial pain, 10% of intracranial tumors and an average of 6.3 years until diagnosis; while Carbamazepine was temporarily helpful, tumour surgery was usually very effective.

Tinnitus is still a rather neglected symptom. 58% of these meningioma patients suffered tinnitus for 3 years. On evaluation of tinnitus patients, Lechtenberg and Shulman [11] found in 21% some responsible neurologic disease and stressed the importance of further diagnostics of this symptom.

Predisposing Factors, Recurrences

Black [1] reviewed the current knowledge of origin and treatment of meningiomas. Factors predisposing to meningioma formation include female sex, previous ionizing radiation, and Type 2 neurofibromatosis. Although meningioma surgery is sometimes thought of as benign and curative, the reported surgical mortality rate is as high as 14.3%. In our patients the prevalence for females with 80% was obvious. There was no direct surgical mortality, but one neurofibromatosis patient died 4 months after surgery. In an analysis of intracranial meningioma recurrences Kajiwara *et al.* [9] found that patients with recurrences were younger than 60 years old, and particularly in the patients who were younger than 40 years the recurrence rate was high. Mean age in the recurrent group was younger and males were more often affected and at shorter intervals. Mirimanoff *et al.* [13] had not found any difference in the "recurrence or progression rates according to the patients' age or sex, or the duration of symptoms".

Surgical Radicality

Most authors report on the specific problems imposed by meningiomas surrounding nerval and vascular structures [8,14] and by the fact of higher rates of recurrence. Therefore some authors postulate "wide excision" [10,14]. Hyperostotic changes of the petrous bone indicate tumorous changes. These cannot be removed totally without sacrificing the labyrinth and/or the cochlea, i.e. hearing and vestibular control. However, Langman postulates that in extensive involvement of the IAC, the inner ear and the deeper portions of the temporal bone, exenteration of the cochlea and semicircular canals should be considered. With regard to the goal of functional neurosurgery we advocate a compromise; as the bony tumorous changes are not life threatening to the patient, it might be useful to leave a functioning auditory system intact as long as possible. The same applies for functional trigeminal or caudal cranial nerves that might continue to work for years though some tumorous infiltration has been found at surgery. However, a paralytic nerve is to be resected and, if possible, reconstructed by sural grafting. In classical presentation of trigeminal neuralgia, the nerve might recover with tumour reduction, in case of severe dysesthesia and atypical pain resection might be necessary.

Auditory Function

In 6% of our patients some improvement of hearing function could be observed after surgery; in a few cases recovery from complete deafness occurred.

Christiansen and Greisen [5] reported on one patient with return of vestibulo-cochlear functions after removal of a CPA meningioma. Maurer and Okawara [12] reported on a case of "cerebellopontine angle meningioma with restoration of hearing from a profoundly deaf state"; he postulates that because of the difficulties in differentiating between CPA meningioma and neurinoma preoperatively and because of the spectacular chances of hearing restoration in meningiomas any destructive surgical approaches must be avoided and the "suboccipital approach maximizes the opportunity for restoration of hearing". Goebel and Vollmer [6] reported on a 41-year-old woman with unilateral severe sensorineural hearing loss and 0% speech discrimination who experienced a remarkable recovery of hearing after removal of a 4 cm meningioma to 86% SDS and stable findings 1 year postoperatively, with no evidence of persistent tumor on repeat MRI scan. Although the pathophysiological changes at the tumour nerve border in neurinomas and meningiomas and the differences in either group are not well understood, the indication for the usefulness of trying to preserve hearing in as many cases as possible, is clear; moreover, the presented cases show the chances of restoration of auditory function also in cases with bad or no clinical preoperative hearing.

Conclusions

CPA meningiomas affect females 4 times more often and at a younger age than males. Hypacusis (72% over 4 years) and vestibular disturbances (61% over 2 years) are the most frequent symptoms and signs. Involvement of the trigeminal nerve (19%), the facial nerve (11%), the caudal cranial nerves (6%) or of the abducens nerve (3%) may precipitate diagnostics. Compared to other CPA lesions unspecific headaches and trigeminal nerve disturbances are especially frequent. The tumour–nerve relationship becomes evident only at surgery where either cranial nerve might be found infiltrated by tumour in 1 to 6%. Complete tumour removal (95%) is less frequently to be achieved than in acoustic neurinomas. Postoperative recovery occurs very early and effectively with regard to vestibular function and to trigeminal function. Early postoperative worsening of facial nerve function in 6 to 10% is followed by a good recovery within some months. Cochlear nerve function is by far less impaired when compared to other CPA lesions, and marked improvement is possible.

References

1. Black PM (1993) Meningiomas. Neurosurgery 32: 643–57
2. Bullitt E, Tew JM, Boyd J (1986) Intracranial tumors in patients with facial pain. J Neurosurg 64: 865–871
3. Catz A, Reider-Groswasser I (1986) Acoustic neurinoma and posterior fossa meningioma. Clinical and CT radiologic findings. Neuroradiology 28: 47–52
4. Cheng TM, Cascino TL, Onofrio BM (1993) Comprehensive study of diagnosis and treatment of trigeminal neuralgia secondary to tumors. Neurology 43: 2298–2302
5. Christiansen CB, Greisen O (1975) Reversible hearing loss in tumours of the cerebello-pontine angle. J Laryngol Otol 89: 1161–1164
6. Goebel JA, Vollmer DG (1993) Hearing improvement after conservative approach for large posterior fossa meningioma. Otolaryngol Head Neck Surg 109: 1025–1029
7. Granick MS, Martuza RL, Parker SW, Ojemann RG, Montgomery WW (1985) Cerebellopontine angle meningiomas: clinical manifestations and diagnosis. Ann Otol Rhinol Laryngol 94: 34–38
8. Javed T, Sekhar LN (1991) Surgical management of clival meningiomas. Acta Neurochir (Wien) [Suppl] 53: 171–182
9. Kajiwara K, Fudaba H, Tsuha M, Ueda H, Mitani T, Nishizaki T, Aoki H (1989) Analysis of recurrences of meningiomas following neurosurgical resection. No Shinkei Geka 17: 1125–1131 (Jpn)
10. Langman AW, Jackler RK, Althaus SR (1990) Meningioma of the internal auditory canal. Am J Otol 11: 201–204
11. Lechtenberg R, Shulman A (1984) The neurologic implications of tinnitus. Arch Neurol 41: 718–721
12. Maurer PK, Okawara SH (1988) Restoration of hearing after removal of cerebellopontine angle meningioma: diagnostic and therapeutic implications. Neurosurgery 22: 573–575
13. Mirimanoff RO, Dosoretz DE, Linggood RM, Ojemann RG, Martuza RL (1985) Meningioma: analysis of recurrence and progression following neurosurgical resection. J Neurosurg 62: 18–24
14. Molony TB, Brackmann DE, Lo WW (1992) Meningiomas of the jugular foramen. Otolaryngol Head Neck Surg 106: 128–136
15. Russel DS, Rubinstein LJ (1989) Pathology of tumours of the nervous system, 5th Ed. Arnold, London
16. Samii M, Draf W (1989) Surgery of the posterior skull base: meningiomas of the cerebellopontine angle. In: Surgery of the skull base. An interdisciplinary approach. Springer, Berlin Heidelberg New York Tokyo, pp 386
17. Samii M, Matthies C (1994) Indication, technique and results of facial nerve reconstruction. Acta Neurochir (Wien) 130: 125–139
18. Tator CH, Duncan EG, Charles D (1990) Comparisons of the clinical and radiological features and surgical management of posterior fossa meningiomas and acoustic neuromas. Can J Neurol Sci 17: 170–176

Correspondence: Dr. Cordula Matthies, Neurochirurgische Klinik, Krankenhaus Nordstadt, Haltenhoffstr. 41, D-30167 Hannover, Federal Republic of Germany.

Acta Neurochir (1996) [Suppl] 65: 92–94
© Springer-Verlag 1996

Management of Petroclival Meningiomas: A Critical Analysis of Surgical Treatment

M. Tatagiba, M. Samii, C. Matthies, and P. Vorkapic

Department of Neurosurgery, Nordstadt Hospital, Hannover, Federal Republic of Germany

Summary

Treatment of petroclival meningiomas have been a matter of discussion in neurosurgery. Since the advent of microsurgery, and with development of new skull base approaches more recently, the treatment of these tumours has become standardised, and the postoperative results considerably improved. However, potential complications have been related with the surgical removal of these lesions. The authors describe their experience and summarise the major reports of the literature on this subject.

Keywords: Approach meningioma; morbidity; petroclival.

Introduction

Petroclival meningiomas represent a difficult task in Neurosurgery, because of the deep localisation of the tumour at the petrous bone and clivus, the involvement of important neurovascular structures, and the tumour extension into the middle cranial fossa, the cerebelloponine angle, the cavernous sinus and the foramen magnum (Fig. 1).

Recent advances in microsurgery and the development of modern skull base approaches have made the surgical resection of these tumours feasible with reduced mortality and morbidity rates [1,5–7]. However, increasing experience has demonstrated that the indication for surgery may be limited in some patients, who do not tolerate an extensive tumour resection without having severe postoperative neurological deficits.

The purpose of this report is to evaluate the surgical series of petroclival meningiomas of the Nordstadt Hospital, Hannover, and to undertake a critical analysis of the literature.

Material and Methods

Between 1978 and 1992, 54 patients with petroclival meningiomas have been operated on in the Neurosurgery Department of Nordstadt Hospital, Hannover. These cases constituted the aim of this study. All patients were examined with CT scan, and in 36 cases MRI was carried out. Four-vessels cerebral angiography and/or MRI-angiography was performed in all cases to study the tumour vascularity, the arterial collateralisation and the venous drainage.

Major reports of the literature on petroclival meningiomas were analysed in order to evaluate the surgical results [1–3,5–8].

Results

There were 39 women and 15 men in our series, and the average age was 47 (17–64) years. Major clinical symptoms at time of admission in the hospital were headache (60%), gait disturbance (55%), and double vision (20%). Major signs were related to cranial nerve deficits (90%), and cerebellar signs (70%). The VIIIth (70%) and the Vth (48%) cranial nerves were the most frequently affected.

Most common surgical approaches used were the combined presigmoid-subtemporal approach (44%), and the retrosigmoid approach (35%). Total tumour removal (Simpson I–II) was achieved in 70% of cases. The most important factors impairing the completeness of surgical removal were tumour invasion into the cavernous sinus, and the infiltration of the pial sheath of major vessels and the brain stem by the tumour.

There was one surgical mortality (2%). Major surgical complications were hemiparesis (15%), CSF leakage (9%), and cranial nerve deficits (13–37%).

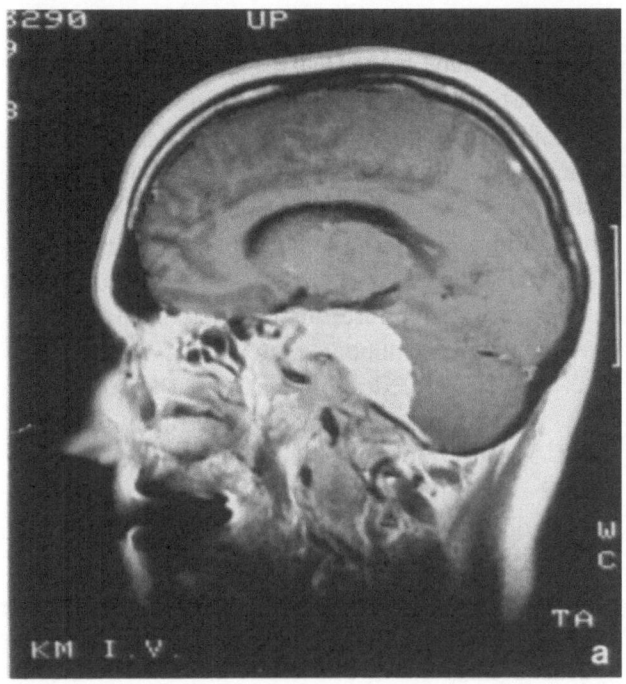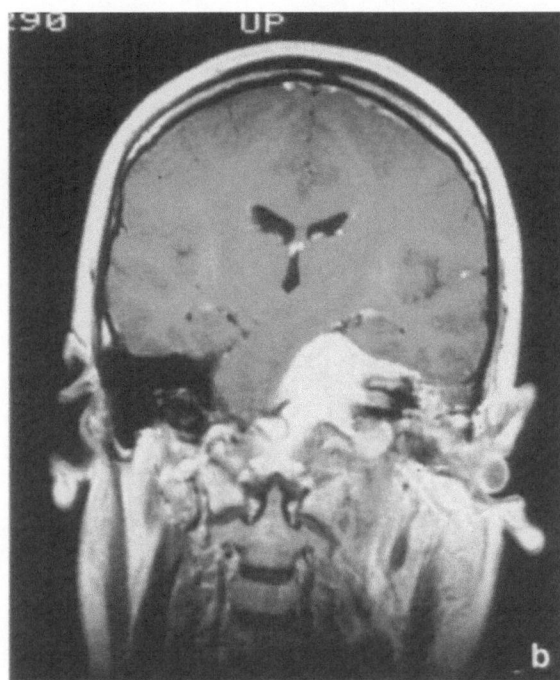

Fig. 1. Gadolinium-enhanced MRI in sagittal (a) and coronal (b) view showing a large petroclival meningioma. The tumour extends into the middle fossa involving the internal carotid artery (a), as well as into the posterior fossa up to foramen magnum displacing the brain stem and growing into the internal auditory canal (b)

Post-operative outcome showed good results in 34 cases (63%), fair in 12 cases (22%), and poor in 8 cases (15%).

Data of the review of the literature are summarized at Tables 1 and 2.

Discussion

Surgical management of meningiomas involving the petroclival region has been considered a difficult challenge in neurosurgery for many years. Petroclival meningiomas originate in the petrous ridge and frequently involve the tentorium, the middle cranial fossa, the cavernous sinus, cerebellopontine angle, the region of foramen magnum, and even the extracranial spaces [6]. Before the microsurgical era, these tumours were usually considered inoperable. High surgical mortality and morbidity were related to the large tumour size and involvement of important neurovascular structures [2,3,8].

In recent years, surgery for petroclival meningiomas has gained increased popularity, and a number of approaches has been described to treat these lesions [5]. Basically, two groups of approaches are recognized: the *anterior approaches*, such as the pterional and the subtemporal routes, and the *posterior approaches*, such as the suboccipital lateral and the presigmoid

Table 1. *Operative Mortality in Early Microsurgical Series*

Author, year	Mortality %
Hakuba, 1977	17%
Yasargil, 1980	15%
Mayberg and Symon, 1986	9%

Table 2. *Major Recent Reported Series on Petroclival Meningioma*

Author, year	No. of cases	Total removal	Mortality	Major morbidity
Yasargil et al., 1980	20	35%	10%	30%
Mayberg and Symon, 1986	35	26%	9%	34%
Hakuba et al., 1988	8	75%	12%	
Al-Mefty et al., 1988	13	85%	0	8%
Sekhar et al., 1990	41	78%	2%	12%
Samii and Tatagiba, 1992	36	75%	0	14%
Spetzler et al., 1992	13	70%	0	15%

routes with or without division of sigmoid sinus. The anterior and the posterior approaches can be combined, originating the *combined presigmoid-subtemporal approach*. The combined presigmoid-subtemporal approach provides an excellent exposure of the petroclival area, the cerebellopontine angle, and the middle cranial fossa (Fig. 2). The surgical technique has been described in details elsewhere.

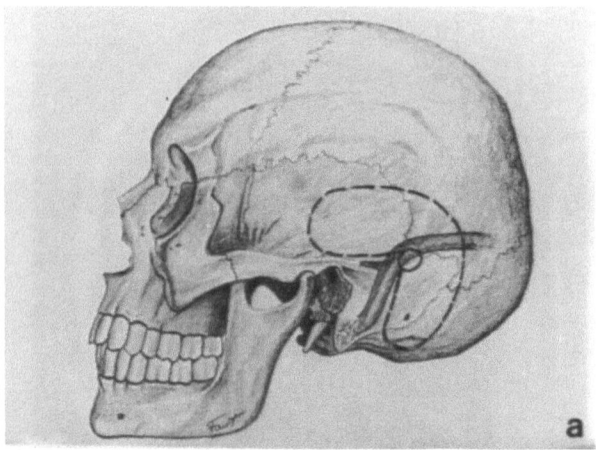

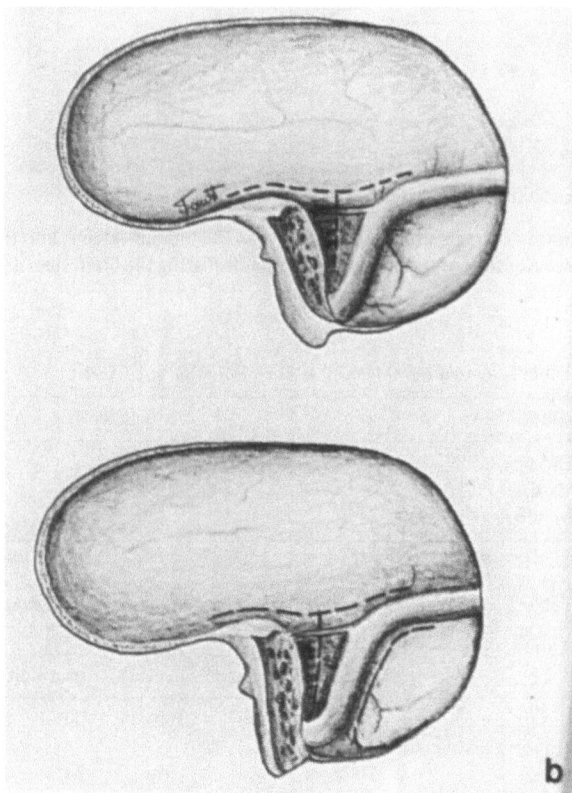

Fig. 2. Drawings of the combined presigmoid-subtemporal approach. The burr hole is placed about the angle between the transverse and sigmoid sinuses, and a craniotomy is carried out over the suboccipital and the subtemporal areas (a). Then, the mastoid is drilled away to expose the dura anterior to the sigmoid sinus (Trautman's triangle) and the superior petrosal sinus (b). The dura is cut in T-fashion. When the tumour extends downward to foramen magnum, a retrosigmoid exposure may be necessary, as demonstrated in (b)

Major complications of petroclival meningiomas surgery include cranial nerve injury, vascular injury and CSF leakage. The anterior approaches may be complicated by injury of the vein of Labbé and the temporal lobe. The posterior approaches may be complicated by lesion to sigmoid sinus, thrombosis of sigmoid and lateral sinuses, injury to the labyrinth block, and CSF leakage.

Review of literature revealed that in most recent series complete tumour removal was achieved in 70% of cases. Indeed, approximately 30% of meningiomas were not resectable. Incomplete tumour removal and surgical morbidity have been largely related to absence of arachnoid plane due to tumour infiltration, which may result in rupture of the pia and pial vessels of the brain stem with consequent brain stem infarction. To be aware of these risks and surgical limitations is essential for management of petroclival meningiomas.

References

1. Al-Mefty O, Fox JL, Smith RR (1988) Petrosal approach for petroclival meningiomas. Neurosurgery 22: 510–517
2. Hakuba A, Nishimura S, Tanaka K, Kishi H, Nakamura T (1977) Clivus meningiomas: six cases of total removal. Neurol Med Chir (Tokyo) 17: 63–77
3. Mayberg MR, Symon L (1986) Meningiomas of the clivus and apical petrous bone: report of 35 cases. J Neurosurg 65: 160–167
4. Samii M, Ammirati M (1988) The combined supra-infratentorial presigmoid sinus avenue to the petro-clival region. Surgical technique and clinical applications. Acta Neurochir (Wien) 95: 6–12
5. Samii M, Tatagiba M (1992) Experience with 36 surgical cases of petroclival meningiomas. Acta Neurochir (Wien) 118: 27–32
6. Sekhar LN, Jannetta PJ, Burkhart RN, Janosky JE (1990) Meningiomas involving the clivus: a six-year experience with 41 patients. Neurosurgery 27: 764–781
7. Spetzler RF, Daspit CP, Pappas CTE (1992) The combined supra- and infratentorial approach for lesions of the petrous and clival regions: experience with 46 cases. J Neurosurg 76: 588
8. Yasargil MG, Mortara RW, Curic M (1980) Meningiomas of basal posterior cranial fossa. Adv Tech Stand Neurosurg 7: 1–115

Correspondence: Dr. med. Marcos Tatagiba, Neurochirurgische Klinik, Krankenhaus Nordstadt, Haltenhoffstr. 41, D-30167 Hannover, Federal Republic Germany.

Acta Neurochir (1996) [Suppl] 65: 95–98
© Springer-Verlag 1996

Symptomatology, Surgical Therapy and Postoperative Results of Sphenoorbital, Intraorbital-Intracanalicular and Optic Sheath Meningiomas

R. Verheggen[1], **E. Markakis**[1], **H. Mühlendyck**[2], and **M. Finkenstaedt**[3]

[1]Clinic of Neurosurgery, [2]Clinic of Neuroopthalmology and [3]Department of Neuroradiology, University of Göttingen, Göttingen, Federal Republic of Germany

Summary

A series of 7 patients with optic sheath meningiomas, 3 intra-canalicular and intraorbital, 2 intraosseus meningiomas of the sphenoid wing involving the optic canal, and 4 sphenoorbital meningiomas were reported. The choice of a surgical approach to the orbit was appropriate to the location and size of the tumour relative to the optic nerve. The most common complaints were proptosis, reduction of visual acuity and paresis of eye muscles. Patients with optic sheath meningiomas are threatened postoperatively by visual loss whereas the high recurrence rate has to be taken into consideration in cases of sphenoorbital meningiomas.

Keywords: Sphenoorbital meningioma; optic canal meningioma; optic sheath meningioma; orbitotomy; visual acuity.

Introduction

Sphenoorbital, optic canal and optic sheath meningiomas have the following symptom complex in common: unilateral protrusion, reduction of visual acuity and paralysis of eye-muscles. Complete surgical excision is the benchmark of successful therapy but requires the earliest possible operation and a proper surgical access depending on the localisation and extension of the tumour. Especially the recurrence rate of sphenoorbital meningiomas, is still estimated between 35% and 50% by the study group of Maroon [8] and Adegbite [1] and has prompted the combination of an intracranial with an intraorbital approach.

Optic sheath meningiomas are often explored by a lateral orbitotomy according to Krönlein [5], but this access is contra-indicated once a tumour reaches the apex or the intracanalicular portion of the orbit [7]. Because of the better anatomical overall view we prefer our modified frontobasal craniotomy and orbitotomy with en bloc resection of the orbital roof especially in cases of retrobulbar or apical tumour localisation.

We intend to appraise the advantages and risks of our preferential surgical procedures as well as to describe complications due to particular topographic conditions from the cases presented.

Patients and Methods

Between January 1987 and April 1994, 46 patients with intra-orbital tumours underwent a frontobasal osteoplastic orbitotomy. Seven of these patients suffered from optic sheath meningiomas and in three other cases intracanalicular and intraorbital meningiomas were diagnosed. Both groups were treated by an orbitotomy. In two cases of intraosseus meningiomas of the sphenoid wing involving the optic canal, we combined a frontopterional access with an orbitotomy. In addition, a frontopterional access with an enlargement of the optic canal was chosen in four cases of sphenoorbital meningiomas.

Radiological Investigation

Besides plain skull roentgenograms to demonstrate a hyperostotic thickening, the initial neuroradiological investigation for all patients included transversal and coronal CT scans with multi-planar reconstructions. The dynamic CT (DCT) was of high diagnostic value in the analysis of perfusion patterns especially in intra-osseus and sphenoorbital meningiomas. The typical feature of meningiomas in DCT is the incipient hyper-vascular phase within the tumour followed by an elevated plateau as described in detail by Jinkins and Sener [4]. Axial T1- and T2-weighted MR-images and sagittal scans before and after application of Gadolinium DTPA were planned in 11 of 16 patients presented in this paper.

Surgical Approaches

A panoramic orbitotomy with frontobasal craniotomy is recommended in cases of retrobulbar or apical tumour localisation [9]. Hence in cases of optic sheath or intraorbital meningiomas we favour a frontobasal osteoplastic orbitotomy. The patient's head was fixed in a sugita clamp under general anaesthesia and in supine position after reclination of about 30° and rotation to the opposite side of about 15°. Then a trapezoid shaped frontobasal craniotomy was carried out attaining the supra orbital margin of the orbit. After mild retraction of the frontal lobe, the orbital roof and supra orbital margin up to the orbital tip was removed en bloc and the tumour resected. Finally the orbital roof was reconstructed using the micro fragmentation plates evolved by Luhr [6]. The sphenoorbital and intracanalicular meningiomas were approached by the classic pterional craniotomy and combined either with the above mentioned orbitotomy or an enlargement of the optic canal. In cases of a distinct extradural tumour

expansion, we avoided incision of the dural sheath of optic nerve otherwise we prefer to expose the vagina externa nervi optici as described by Dolenc [3].

Representative Case Reports

Case 1

An eighty-two-year-old patient complained about a reduction of visual acuity of the right eye and the projection of shadows. The CT examination with multiplanar reconstructions exposed a retrobulbar and apical localised process of 1.5 cm in diameter (Fig. 1). Because of risks relating to her age and general condition, an operation was refused. But after a rapid deterioration of vision to the point of amaurosis and fearing a further intracranial tumour invasion, the woman decided to undergo a surgical intervention. The optic sheath meningioma was resected by a frontobasal orbitotomy. As expected, the vision remained unchanged.

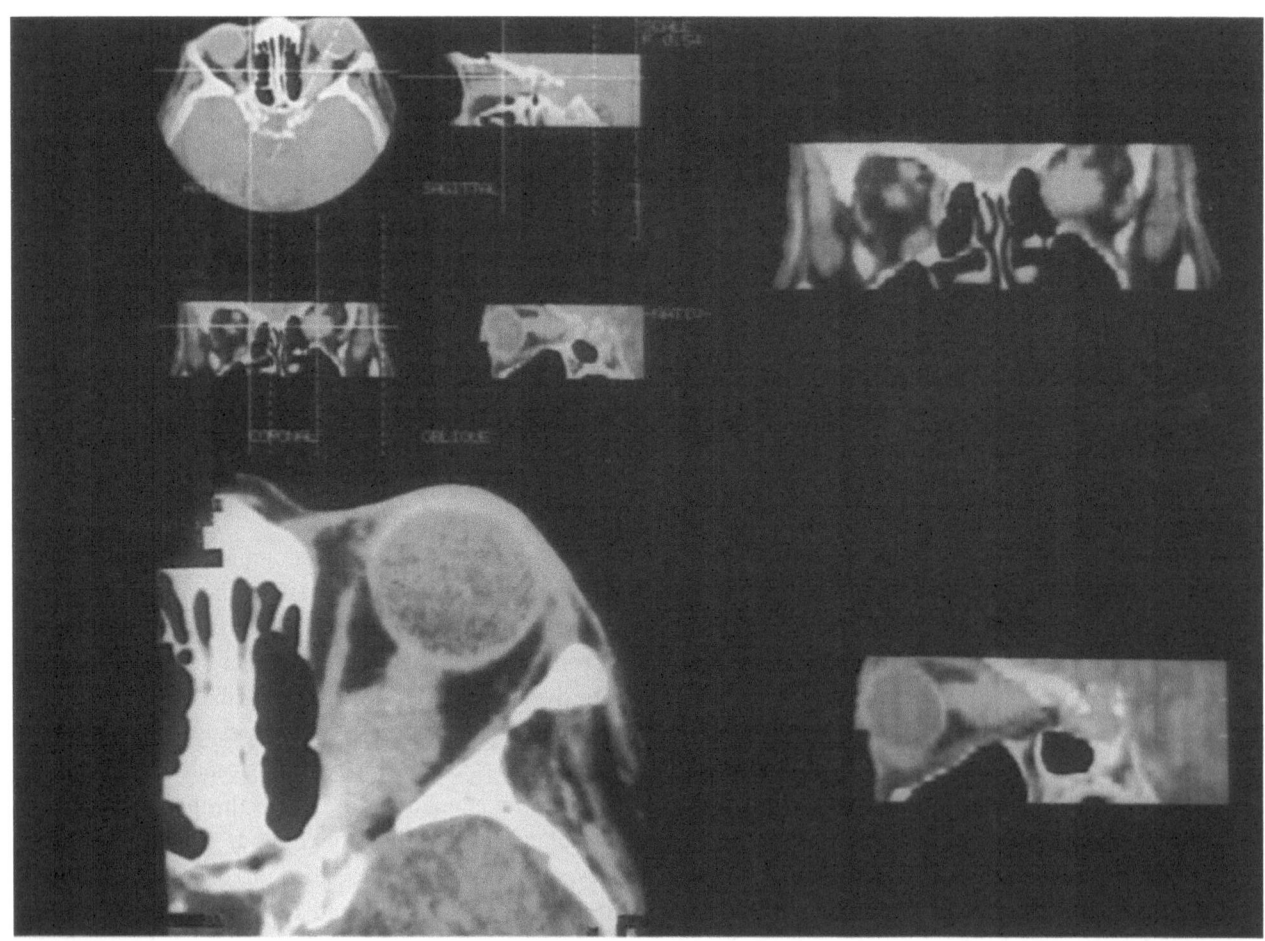

Fig. 1. Case 1: preoperative sagittal, coronal and axial CT scans and reconstructions revealing a retrobulbar and apically localised optic sheath meningioma

Case 2

In October 1992, a fifty-five year-old woman noticed a progressive exophthalmus. The neuroophthalmological examination revealed a protrusio bulbi of 6–7 mm. The CT examination with osseous reconstructions exposed a marked hyperostosis of the sphenoid bone which was first misdiagnosed as fibrous dysplasia. After application of contrast media the MRI confirmed a distension of the dura. The typical perfusion pattern of the DCT already led preoperatively to the diagnosis of an en plaque growing meningioma (Fig. 2a–c). A pterional craniotomy was combined with an orbitotomy and an enlargement of the optic canal. Subsequently the minor wing of sphenoid bone and the lateral wall of the orbital cavity were partially resected. The intracranial portion of the meningioma was exstirpated and the dural sheath of optic nerve was incised. The lateral wall of the orbit was reconstructed by a split calvarian cranioplasty. Postoperatively the protrusion declined rapidly.

Results

In 89% of the patients with sphenoorbital (4), intraosseus and sphenoid (2) and intracanalicular and intraorbital meningiomas (3) postoperative follow-up investigation by neuroophthalmologists revealed an improvement of visual acuity, decline of protrusion and an amelioration of eye muscles paresis. In 60% of our patients with frontobasal orbitotomy a temporary paresis of the upper eyelid was observed, which however, improved spontaneously within 6–8 weeks.

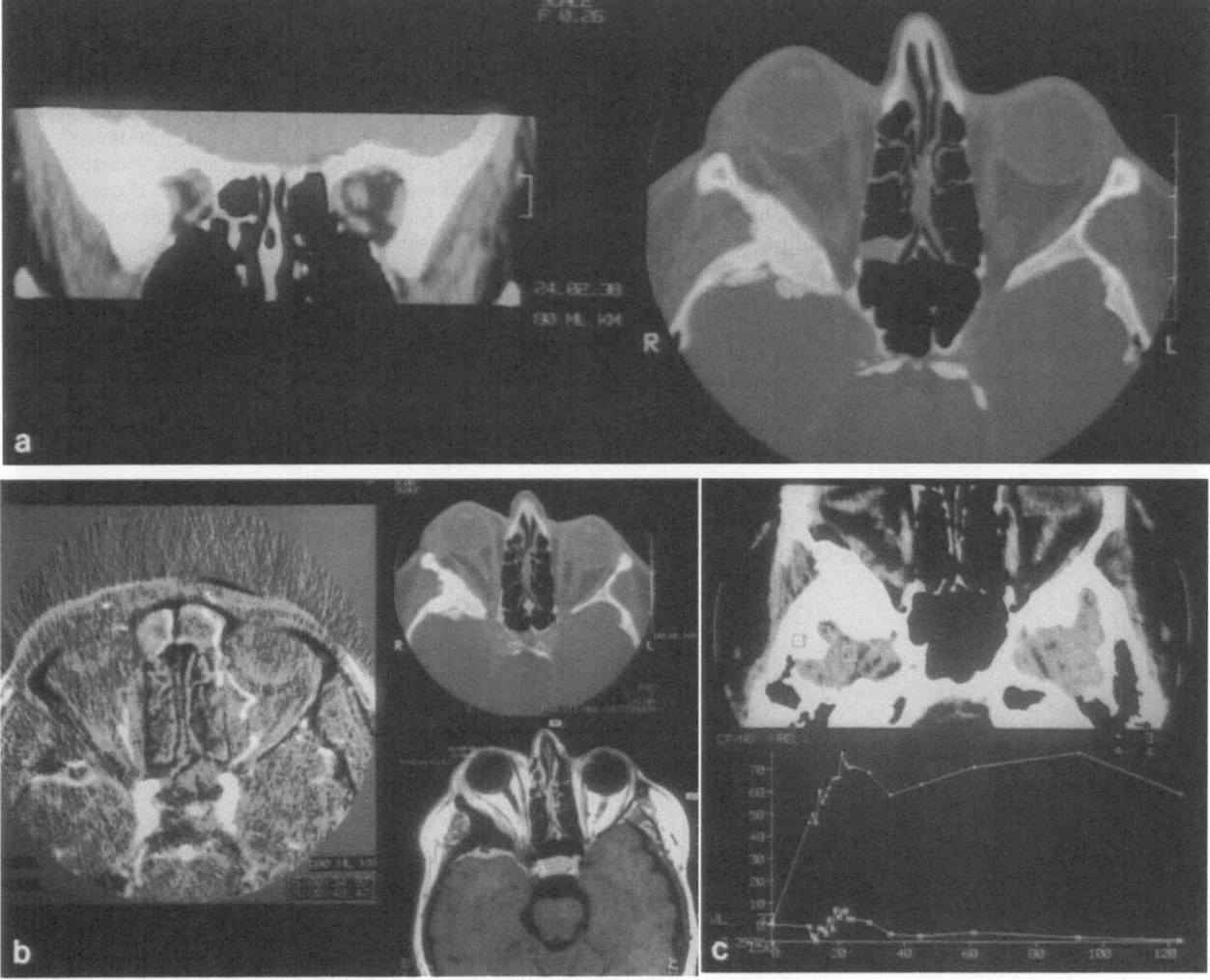

Fig. 2. (a) Case 2: CT examination exposing a marked hyperostosis of the sphenoid bone. (b) Case 2: In comparison to the CT scans with osseus reconstructions the T1 weighted MR images with Gadolinium DTPA and the angioMRI confirmed a distension of the dura. (c) Case 2: Dynamic CT with the typical perfusion pattern: the initial hypervascular phase is followed by a nearly horizontal and elevated plateau

The prognosis of patients with optic sheath meningiomas is dependent on the time of the diagnosis, tumour localisation and intraoperative complications. Unfortunately, an amaurosis was existent in three patients prior to surgery. Although vision was reduced in three further cases postoperative re-examinations verified an unchanged or slightly improved vision. In one patient with an optic sheath meningioma, a unilateral amaurosis arose due to injury of the ophthalmic artery.

Discussion

A surgical approach to the orbit relies on the location of the tumour, the size and the vascularity of the lesion. Due to their different origin, topography, and prognosis orbital meningiomas are subdivided either in primary or secondary tumours [2,10]. The primary meningiomas arise either from the gap cells of the arachnoid surrounding the optic nerve or from the cells external to the dura but within the orbital periosteum. Secondary meningiomas extend into the orbit from adjacent structures by expansion of the sphenoid wing, tuberculum sellae or the anterior clinoid process. Therefore, the predominant feature of optic sheath meningiomas is early visual loss. This is in contrast to secondary meningiomas, which – with the exception of apically localised tumours – cause proptosis before signs of nerve compression [10].

According to our experiences, early surgery improves the chance of preserving visual acuity in cases of optic sheaths meningiomas with an extradural extension and/or located in the anterior or middle third of the optic nerve. However, reduction of visual acuity is often unavoidable in apically located tumours due to the complex structure of vessels, cranial nerves and ocular muscles in the orbital funnel (tendon of Zinn). Although there is a minor risk to jeopardise visual acuity in secondary meningiomas during surgery, the high recurrence rate of sphenoorbital tumours requires close-meshed follow-ups.

Acknowledgements

The authors thank Dr. G. Latta, Mrs. R. Vania and J. Donohoe for the critical revision of the manuscript.

References

1. Adgebite AB, Kahn MI, Paine KWE, Tan LK (1983) The recurrence of intracranial meningiomas after surgical treatment. J Neurosurg 58: 51–56
2. Craig W, Gogela LJ (1949) Intraorbital meningiomas. A clinicopathological study. Am J Ophthalmol 32: 1663–1680
3. Dolenc VV (1989) Anatomy and surgery of the cavernous sinus. Springer, Heidelberg New York Tokyo
4. Jinkins JR, Sener RN (1991) The characteristics of cerebral meningiomas and surrounding tissues on dynamic CT. Neuroradiology 33: 499–506
5. Krönlein RU (1988) Zur Pathologie und operativen Behandlung der Dermoidzysten der Orbita. Beitr Klin Chir 4: 149–163
6. Luhr HG (1987) Versorgung begleitender Frakturen bei Gesichtsverletzungen. Langenbecks Arch Chir 372: 687–695
7. Maroon JC, Kennerdell JS (1984) Surgical approaches to the orbit. J Neurosurg 60: 1226–1235
8. Maroon JC, Kennerdell JS, Vidovich DV, Abla A, Sternau L (1994) Recurrent sphenoorbital meningioma. J Neurosurg 80: 202–208
9. Rootman J (1988) Orbital surgery. In: Rootman J (ed) Diseases of the orbit – a multidisciplinary approach. Lippincott, Philadelphia, pp 579–611
10. Wright JE, Call NB, Liaricos S (1980) Primary optic nerve meningioma. Br J Ophthalmol 64: 533–558

Correspondence: R. Verheggen, M.D., Clinic of Neurosurgery, University of Göttingen, Robert-Koch-Str. 40, D-37075 Göttingen, Federal Republic of Germany.

Acta Neurochir (1996) [Suppl] 65: 99–101
© Springer-Verlag 1996

Factors Influencing Morbidity and Mortality After Cranial Meningioma Surgery – a Multivariate Analysis

J. Meixensberger, T. Meister, M. Janka, B. Haubitz, K.A. Bushe, and **K. Roosen**

Neurochirurgische Klinik, Poliklinik, and Rechenzentrum, Universität Würzburg, Federal Republic of Germany

Summary

In a retrospective analysis 385 patients with a histologically defined cranial meningioma were studied to analyze the impact of characteristic factors on morbidity and mortality after modern cranial meningioma surgery. Mortality was 4.2% one month and 7.3% six months after operation. 15.6% of the patients stayed more than one month in the hospital (defined as criteria of operative morbidity). Age, poor preoperative clinical condition (ASA score), intra- and postoperative bleeding and CSF disturbances were significantly associated with a subsequent decrease of quality of life. First symptoms like intracranial hypertension, seizures, aphasia and hemiparesis were correlated with an increase of postoperative Karnowsky index. Postoperative quality of life decreased in patients with optic and other cranial nerve disturbances significantly. Tumour size, location (exception: medial sphenoid wing) and histological diagnosis did not influence surgical outcome. This information may be useful in management decisions regarding asymptomatic meningiomas in elderly and high risk patients.

Keywords: Meningiomas; morbidity; mortality; risk factors.

Introduction

Increased life expectancy [7] together with the availability of safe and routine use of computed tomographic (CT) scanning and magnetic resonance imaging (MRI) are the important reasons of increased incidence of cranial meningiomas today. In the same way the number of patients without or only minimal neurological deficits increased especially in patients aged over 65. Nevertheless microsurgical operation procedures improved and there are recent advances in neuroanesthesia and intensive care management. Performance of major neurosurgical operations in elderly patients may be a higher risk than in younger ones. Therefore it is necessary to evaluate precise predictors of surgical risks in the discussion about the benefits of operating on meningiomas. It was the purpose of this study to analyze the impact of characteristic factors on morbidity and mortality on outcome of meningioma patients after surgery.

Patients and Methods

In a retrospective analysis 385 patients with histological defined cranial meningioma were investigated. All patients were operated consecutively between 1975 and 1988 in the Department of Neurosurgery, University Würzburg. A total of 120 variables of each patient including preoperative status (e.g. age, sex, risk factors, neurological signs, tumour location and tumour size), intraoperative and postoperative course (amount of tumour resection, complications as brain swelling, bleeding, neurological worsening etc.) were documented. Preoperative risk factors were classified using the ASA classification according to the American Society of Anesthesiology [6]. The quality of life was graded using the Karnofsky index (KI) before (K1), 30 days (K2) and six months (K3) postoperatively [3]. At least the increase of Karnofsky index (Kd) was calculated as the difference of Karnofsky index before operation (K1) minus Karnofsky index 6 months after operation (K3). In addition to descriptive statistical procedures (mean, standard deviation) various statistical approaches were used to find the best predictors. Bi- (Mann–Whitney-U-Test, Spearman rank-correlation, Kendalls Tau correlation) and multivariate statistical models (Kruskal–Wallis-, Puri and Sen-test) were used to calculate the influence of each variable on postoperative Karnofsky index. At least the stepwise multiple logistic regression analysis was used to predict the influence on operative outcome of the single variables. P values were significant $p < 0.05*$, $p < 0.01**$ and $p < 0.001***$. Statistical analysis and graphical output were generated with the following statistical software packages: MEDAS (Copyright Rechenzentrum Würzburg) and HARVARD GRAPHICS (Copyright Software Publishing Corporation).

Results

Demographic, Clinical, Neuroradiological and Histological Data

Ratio of sex was 128 male to 257 female patients. Mean age at operation was 53.1 years (+/–13.2). One third of the patients was older than 60 years. According

Table 1. *Dependency of First Clinical Symptom on Increase of Karnofsky Index 6 Months Postoperatively*

First symptom	Increase of Kd Mean +/– SD	P	N
Aphasia	15.0 +/– 14.6	**	5
Hemiparesis	10.7 +/– 25.2	**	28
Personality changes, dementia	9.8 +/– 22.5	**	30
Disturbances of gait	7.1 +/– 31.3	**	12
Anosmia	4.2 +/– 13.6	**	6
Headache	3.9 +/– 29.1	**	89
Seizure	3.8 +/– 16.4	**	81
Visual deficit	–2.7 +/– 24.7	**	28
Cranial nerve deficit	–3.9 +/– 29.0	**	23
Exophthalmus	–4.0 +/– 24.4	**	20

Increase of Karnofsky index (Kd) = Karnofsky index 6 months postoperatively minus Karnofsky Index preoperatively. Key to P values $p < 0.01$**. *N* number of patients.

to the ASA classification 5.5% of the patients were graded ASA I, 41.5% ASA II, 49.2% ASA III and 3.8% ASA IV. The first clinical symptoms and signs are listed in Table 1. Headaches, seizures and mental disturbances headed the list. The meningiomas were located as follows: convexity (34.4%), sphenoid wing, tuberculum sellae (19%), base of skull (16.2%), falx (16%), tentorium (6.1%), petrous ridge (6.1%) and others (2.2%). Computer tomographic calculated mean tumour size was 40.3 mm (+/– 16.4). Almost all meningiomas were histologically benign (endotheliomatous 41.3%, fibrous 26.0%, transitional 23.5%, anaplastic 1.6%, others 7.6%).

Factors Influencing Operative Mortality and Morbidity

The operative mortality was 4.2% one month and 7.3% six months after operation. 15.6% of the patients were admitted more than one month to the hospital (defined as criteria of operative morbidity). Analyzing the postoperative Karnofsky index (K3 and Kd) showed no significant difference between male and female patients. Age as well as preoperative risk factors influenced postoperative outcome. Absolute Karnofsky index decreased in older patients (K1: Rho = –0.3369, K2: Rho = –02829, K3: Rho = –0.3403 $p < 0.001$***). However analyzing Kd of various age groups there was no significant difference between the younger and the older patients (Fig. 1). Figure 2 showed the Karnofsky index of the four ASA groups before and after operation (K1–K3). ASA IV classified patients revealed the lowest pre- and postoperative Karnofsky values. Especially cardiac problems worsen significantly the postoperative quality of life (patient

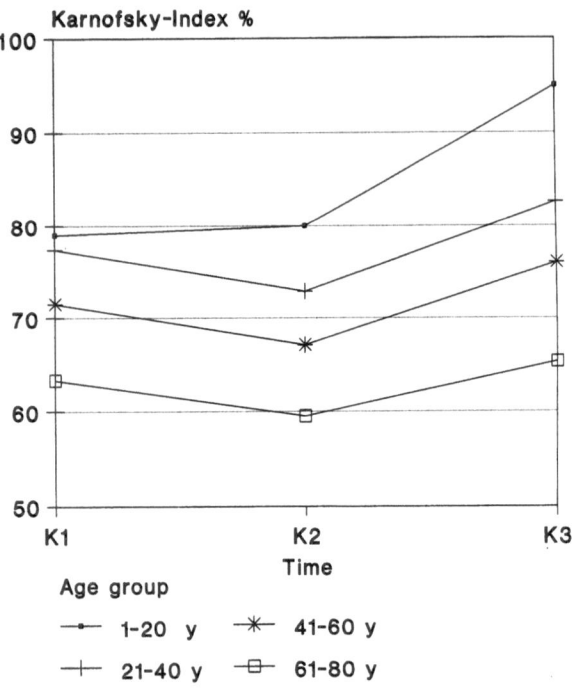

Fig. 1. Karnofsky index of different age groups at various times after surgery. K1 preoperatively, K2 30 days after operation, K3 6 months after operation. y years

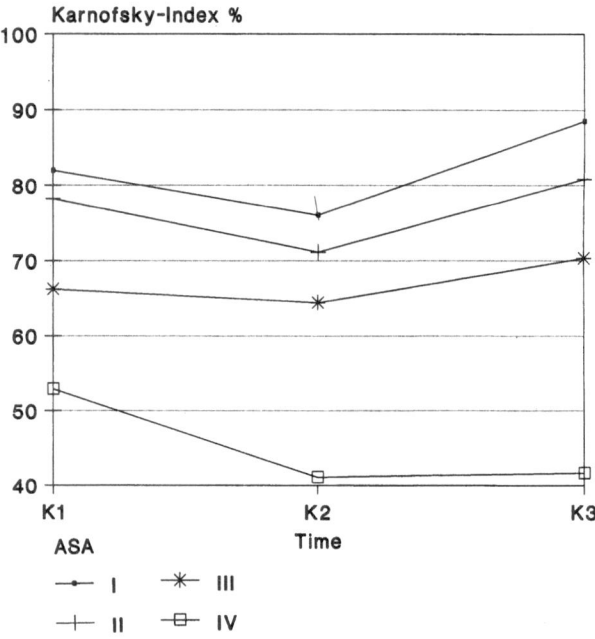

Fig. 2. Karnofsky index of different ASA groups at various times after surgery [6]. K1 preoperatively, K2 30 days after operation, K3 6 months after operation

with cardiac disease (n=75): Kd = –3.7 +/– 30.4, without cardiac illness (n=253): Kd = –6.0 +/– 22.1, $p < 0.05$*). Additionally postoperative Karnofsky index decreased in patients with intra- and postoperative bleeding and disturbances of cerebrospinal fluid circu-

Table 2. *Dependency of Pre- and Intraoperative Target Variables on Increase of Karnofsky Index (Kd) After Operation Using Multiple Regression Analysis*

Kd =	13.3790	
	− 0.5283	per percent of preoperative Karnofsky index (K1)
	− 0.0034	per ml intraoperative blood transfusion
	− 6.6475	preoperative heart failure, cardiac illness
	− 12.5768	intraoperative arterial bleeding
	− 6.6322	medial sphenoidal ridge meningioma
	− 10.3910	disturbances of CSF circulation
	− 0.2719	per year of life
	− 8.3959	preoperative worsening of vision

Kd Karnofsky Index 6 months postoperatively minus Karnofsky index preoperatively.

lation (Rho = −0.01508, $p < 0.05$*). Duration of clinical symptoms was shorter than one year in almost 50% of the patients (mean 25 +/− 35.3 months). There was no correlation to an increase of postoperative Karnofsky index. First symptoms as aphasia, hemiparesis and mental deficits improved after operation and were significantly correlated with an increase of quality of life ($p < 0.001$ ***) (Table 1). But postoperative outcome decreased in patients with optic nerve and other cranial nerve disturbances ($p < 0.001$***). Tumour size, location (medial sphenoidal ridge meningiomas excluded) and histological diagnosis did not affect surgical prognosis after six months. Table 2 summarized the results of stepwise multiple regression analysis and pointed out the strongest predicting variables of postoperative prognosis after meningioma surgery. The increase of Karnofsky index (Kd) was calculated with 13.3790 and e.g. every percent of K1 decreased Kd by the factor 0.5283.

Discussion

Intracranial meningiomas are in more than 90% of the cases histologically benign tumours and usually amenable to surgical treatment. Without any doubt operations on meningiomas are absolutely indicated in patients with intracranial hypertension and major neurological deficits. However CT scanning and MRI have made the incidental diagnosis of a meningioma tumour more frequent [5] almostly in the ageing population. Consequently a critical analysis of operative risks versus the risk of the natural course (growth rate) is necessary to answer the question of benefit of doing an operation on meningiomas.

In our series the overall mortality accounts for 4.2 percent and underlined the improvements in surgical treatment compared to previous studies (mortality range from 4% to 16%) [1,2]. In agreement with other authors, morbidity (15.6%) was influenced by preoperative KI, age, number of risk factors and ASA score pointed out by regression analysis [2,4].

Although age is a predictor of outcome, additional risk factors and poor clinical conditions mainly determine outcome of patients [4]. Therefore age itself is no contraindication for surgery.

Additional operative difficulties related to location (e.g. medial sphenoidal wing), vascularization of the meningioma and disturbances of cerebrospinal fluid circulation should be considered in making operative decision. Sometimes incomplete tumour removal may be useful to prevent intraoperative and estimate postoperative complications (e.g. cranial nerve deficits, ischemia). In summary a balanced strategy is necessary in the treatment of asymptomatic meningiomas of elderly and high risk patients.

References

1. Chan RC, Thomson GB (1984) Morbidity, mortality and quality of life following surgery for intracranial meningiomas: A retrospective study in 257 cases. J Neurosurg 60: 52–60
2. Cornu Ph, Chatellier G, Dagreou F, Clemenceau S, Foncin JF, Rivierez M, Philippon J (1990) Intracranial meningiomas in elderly patients: postoperative morbidity and mortality: factors predictive of outcome. Acta Neurochir (Wien) 108: 98–102
3. Karnofsky DA, Abelmann WH, Craver LF, Burchenal JH (1948) The use of nitrogen mustards in the palliative treatment of carcinoma, with particular reference to bronchogenic carcinoma. Cancer 1: 634–656
4. Maurice-Williams RS, Kitchen ND (1992) Intracranial tumours in the elderly: the effect of age on the outcome of the first time surgery for meningiomas. Br J Neurosurg 6: 131–137
5. Rausing A, Ybo W, Stenflo J (1970) Intracranial meningioma: A population study of ten years. Acta Neurol Scand 46: 102–110
6. Schneider AJL (1983) Assessment of risk factors and surgical outcome. Surg Clin North Am 63: 1113–1119
7. Walker E, Robins M, Weinfeld FD (1985) Epidemiology of brain tumours: The national survey of intracranial neoplasms. Neurology 35: 219–226

Correspondence: PD Dr. J. Meixensberger, Neurochirurgische Klinik und Poliklinik, Universität Würzburg, Josef-Schneider-Strasse 11, D-97080 Würzburg, Federal Republic of Germany.

Acta Neurochir (1996) [Suppl] 65: 102–104
© Springer-Verlag 1996

Clinical Relevance of Somatostatin Receptor Scintigraphy in Patients with Skull Base Tumours

C. Luyken[1], G. Hildebrandt[1], B. Krisch[3], K. Scheidhauer[2], and N. Klug[1]

[1]Departments of Neurosurgery and [2]Nuclear Medicine, University of Köln, Köln, and [3]Department of Anatomy, University of Kiel, Kiel, Federal Republic of Germany

Summary

The differential diagnosis of tumours in the skull base is often difficult. With the experience that various intracranial tumours differ in their expression of somatostatin binding sites (SBS) somatostatin receptor scintigraphy (SRS) with the somatostatin analogue octreotide can give additional information of the tumour entity.

Seventy patients with various tumours of the skull base were examined with [111]Indium-labelled DTPA-octreotide injected i.v. . Planar and tomographic images were obtained with a gamma camera 4–6 and 24 hours after injection. All of the meningiomas (unifocal and multifocal tumours in various locations) showed a high density of SBS whereas in none of the examined neurinomas SR were found. Pituitary adenomas revealed in only 50% SR in different concentrations and independent of the endocrine activity.

SRS can help in the differential diagnosis between meningiomas and other tumours, postoperative scar or radionecrosis at the skull base. A dural infiltration with meningioma tissue ("meningeal sign") may be discriminated from a reactive hypervascularisation in lesions with a diameter > 0.5 cm. We conclude that SRS can offer additional diagnostic aspects in the pre- and postoperative management of patients with skull base tumours.

Keywords: Skull base tumours; somatostatin receptors; in vivo and in vitro; receptor scintigraphy.

Introduction

The variety of tumours located in the skull base need different therapeutical proceedings but the differential diagnosis guided by MRI or CT-scans often remains uncertain. One of the additional discriminating features is the specific existence of somatostatin receptors (SR) in intracranial tumours [1,4–8], which can be examined preoperatively since Krenning *et al.* [2] have developed the SR-scintigraphy (SRS) in vivo using the isotope-labelled somatostatin analogue [111]Indium DTPA-octreotide.

The aim of the study was to prove the diagnostic value in pre- and postoperative management of patients with skull base tumours.

Patients and Methods

Seventy patients with various tumours of the skull base (meningioma: n=39, pituitary adenoma: n=18, neurinoma: n=5, neurofibroma: n=2, lymphoma: n=2, glioma: n=2, epidermoid cyst: n=1, haemangiopericytoma: n=1) shown in MRI or CT scans were examined (Fig. 1). The somatostatin analogue octreotide labelled with [111]Indium-DTPA was injected i.v. as a 10 or 20 μg bolus corresponding to 110 or 220 MBq. Gamma-camera images and SPECT (detection limit =0.5 cm of tumour size) were obtained 3–6 h and 24 h post injection. The scintigraphic evaluation was performed without knowledge of the CT and MRI findings. The histological classification corresponded to the WHO grading system. SR were detected in vitro using somatostatin-gold conjugates[3].

Results

All patients with meningiomas independent of the tumour localisation and classification showed a high focal tracer uptake (Table 1, Figs. 2a–d, 3a,b, 4a,b). In two cases of a meningeal sign in MRI (meningioma infiltration of the dura) only one revealed a focal tracer uptake corresponding to the histological findings (Fig. 2c,d). The other case had no focal tracer uptake and no tumour cells in projection of the meningeal sign in MRI (Fig. 2a,b).

Only in 50% of the pituitary adenomas independent of the hormonal activity an enhanced tracer uptake was detected but with varying intensity (Table 1). In two patients with high focal tracer uptake in the sella no distinction from a meningioma was possible.

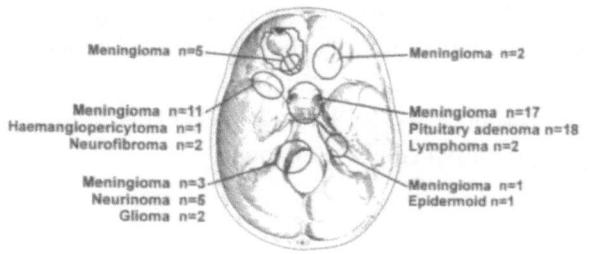

Fig. 1. Outlines the numbers of the examined tumours located at the skull base

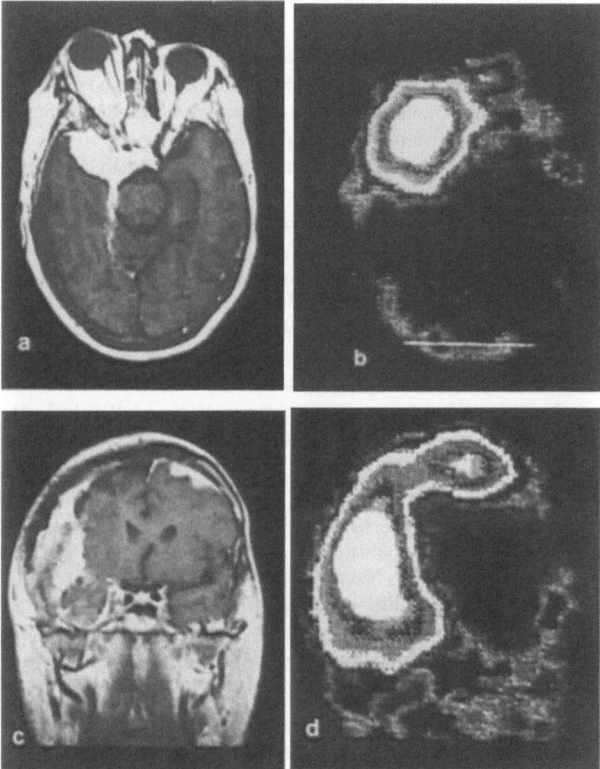

Fig. 2.(a–d) Demonstrates the findings in two patients with recurrent meningiomas. (a) Axial view of a contrast-enhanced MRI shows the hyperintense tumour located to the right sphenoid bone with a suspected dural infiltration. (b) Axial SPECT image after application of ^{111}In DTPA-octreotide reveals a focal tracer uptake located at the right sphenoid bone but not at the site of the suspected dura infiltration. (c) Represents a right frontobasal hyperintense lesion with a dural tumour spread on the left side in a coronal view of a contrast enhanced MRI. (d) Anterior view of a SPECT after somatostatin receptor scintigraphy exposes a high focal tracer uptake in the suspected tumour areas and even in the dura on the left side

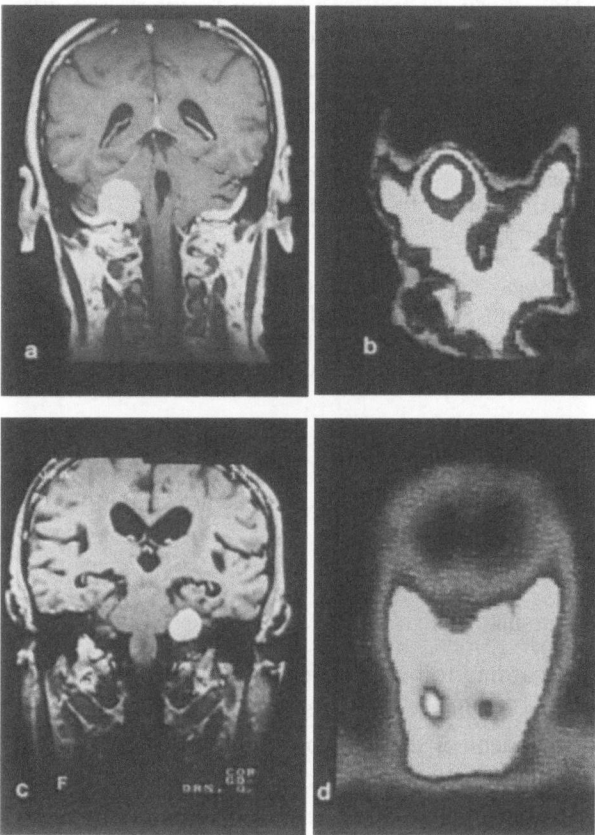

Fig. 3.(a–d) Represents the findings of two patients with tumours in the cerebellopontine angle. (a) Shows a hyperintense tumour with contrast enhancement in a coronal MRI. (b) Reveals a SPECT in a coronal view after radio-labelled somatostatin application i.v., which shows enhanced tracer uptake located at the right cerebellopontine angle. (c) Outlines a contrast-enhanced tumour in the left cerebellopontine angle on the coronal view of MRI. (d) Indicates no enhanced tracer uptake in the cerebellopontine angle after application of labelled octreotide on the coronal view of a SPECT image

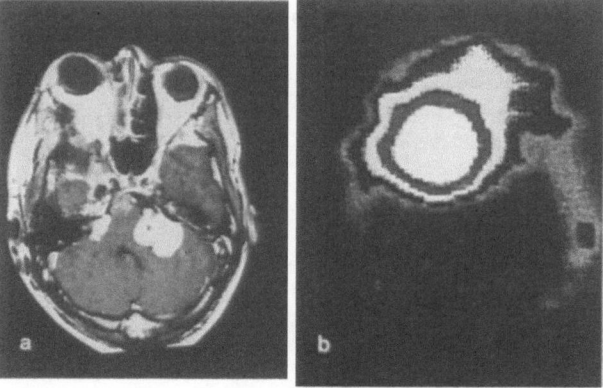

Fig. 4.(a,b) Displays a case of neurofibromatosis II with multiple intracranial tumours. (a) Demonstrates the axial contrast enhanced MRI with a recurrent meningioma on the right sphenoid bone, contrast enhanced tumours in both cerebellopontine angles and a prepontine tumour without contrast enhancement. (b) Axial view of a SPECT image after labelled somatostatin was injected i.v. reveals enhanced focal tracer uptake in the right sphenoid bone but no enhancement in the cerebellopontine angles or in the prepontine region

Five patients with neurinomas did not show somatostatin binding sites in vivo or in vitro (Table 1, Fig. 3c,d).

In a patient with a neurofibromatosis II and multiple intracranial tumours the recurrent meningioma of the right sphenoid bone revealed a high focal tracer uptake

Table 1. *Results of Somatostatin Receptor Scintigraphy with ¹¹¹In-labelled Octreotide in 70 Patients with Skull Base Tumours*

Histology	n	In vivo detection of somatostatin receptors	In vivo receptor density
Meningioma	39	39	+++
Pituitary adenoma	18	8	+−+++
Neurinoma	5	0	0
Neurofibroma	2	2	+
Lymphoma	2	1	0−++
Epidermoid	1	0	0
Haemangiopericytoma	1	1	++
Glioma grade 2 + 2–3	2	0	0

whereas the tumours of the cerebellopontine angle and the glioma in front of the pons demonstrated no enhanced tracer uptake (Fig. 4a,b).

Discussion

According to our findings all patients with meningiomas at the skull base showed a high SR-density independent of the histological subclassification and could be discriminated from neurinomas without SR in the cerebellopontine angle or from radionecrosis or postoperative scar. In cases of a suspected meningioma infiltration in MRI, SRS may differentiate between meningioma tissue and a unspecific hypervascularisation when the suspicious area is not under 0.5 cm.

According to Lamberts *et al.* [3,4] patients with pituitary adenomas only in 50% demonstrated SR in different concentrations and independent of their hormonal activity. For the differentiation of tumours located in the sella region SRS can only exclude a meningioma if no SR are found.

In multiple tumours as in a neurofibromatosis II SRS may help to distinguish the lesions by means of the different SR-density.

We conclude that the SRS might help in the differential diagnosis of skull base tumours.

References

1. Hildebrandt G, Scheidhauer K, Luyken C, Schicha H, Klug N, Dahms P, Krisch B (1994) High sensitivity of the in vivo detection of somatostatin receptors by ¹¹¹Indium (DTPA-octreotide)-scintigraphy in meningioma patients. Acta Neurochir (Wien) 126: 63–71
2. Krenning EP, Bakker WH, Breemann WAP, Koper JW, Kooij PPM, Ausema L, Lameris JS, Reubi JC, Lamberts SWJ (1989) Localisation of endocrine-related tumours with radioiodinated analogue of somatostatin. Lancet i: 242–244
3. Krisch B, Buchholz C, Mentlein R (1991) Somatostatin binding sites on rat diencephalic astrocytes. Lightmicroscopic study in vitro and in vivo. Cell Tissue Res 263: 253–263
4. Lamberts SWJ, Krenning EP, Reubi JC (1991) The role of somatostatin and its analogs in the diagnosis and treatment of tumours. Endocrine Rev 12: 450–482
5. Lamberts SWJ, Koper J, Reubi JC, Krenning EP (1992) Endocrine aspects of diagnosis and treatment of primary brain tumours. Clin Endocrinol 37: 1–10
6. Luyken C, Hildebrandt G, Scheidhauer K, Krisch B, Schicha H, Klug N (1994) ¹¹¹Indium (DTPA-octreotide) scintigraphy in patients with cerebral gliomas. Acta Neurochir (Wien) 127: 60–64
7. Reubi JC, Cortes R, Maurer R, Probst A, Palacios JM (1986) Distribution of somatostatin receptors in human brain: an autoradiographic study. Neuroscience 18: 329–346
8. Scheidhauer K, Hildebrandt G, Luyken C, Schomäcker K, Klug N, Schicha H (1993) Somatostatin receptor scintigraphy in brain tumours. Horm Metabol Res Series [Suppl] 27: 59–62

Correspondence: C. Luyken, M.D., Department of Neurosurgery, University of Köln, Joseph-Stelzmann-Str. 9, D-50931 Köln, Federal Republic of Germany.

Acta Neurochir (1996) [Suppl] 65: 105–107
© Springer-Verlag 1996

Progesterone Receptors in Tumor Fragment Spheroids of Human Meningiomas

J.C. Tonn[1], **M.M. Ott**[2], **W. Paulus**[2], **J. Meixensberger**[1], and **K. Roosen**[1]

[1]Department of Neurosurgery and [2]Institute of Pathology, University of Würzburg, Würzburg, Federal Republic of Germany

Summary

Progesterone receptors (PgR) are detectable in about 60–70% of tissue specimens of human meningiomas. Despite these data, PgR are hardly to be found in monolayer tissue culture of meningiomas. Aim of this study was to elucidate whether PgR might be preserved in tumor fragment spheroids of meningiomas maintained in organ culture since the morphological appearance of the original tumor is preserved by this culture technique.

Aliquots of meningioma specimens of 25 patients (17 females) were snap frozen in liquid nitrogen immediately after removal. Additionally, monolayer tissue cultures of the same specimen were obtained as primary culture and passage #3. Tumor fragment spheroids were kept on medium-agar with liquid medium overlay and harvested after 1 and 3 weeks in culture. PgR were detected by immunohistochemistry using a rat monoclonal antibody.

18/25 meningioma tissue specimens were positive for PgR. In 8 out of 15 PgR-positive tumors which formed spheroids we could detect PgR in fragment spheroids after 1 and 3 weeks in culture. In contrast, none of the monolayers depicted PgR. PgR is preserved in a considerable amount of tumor fragment spheroids of PgR-positive meningiomas. They remain detectable after 3 weeks of culture whereas monolayer tissue cultures are PgR-negative. Thus, tumor fragment spheroids seem to be a suitable tool to investigate progesterone/antiprogesterone effects in vitro.

Keywords: Meningioma; progesterone receptor; cell culture; spheroids.

Introduction

60–70% of all human meningiomas express the nuclear bound progesterone receptor (PgR) as detected by immunohistochemistry [3]. Although some preliminary data seem to show a moderate clinical benefit of antiprogesterone therapy in nonresectable meningiomas these phenomena have not been elucidated in cell culture yet since PgR are not reliably detectable in conventional monolayer culture systems [7,12]. Tumor fragment spheroids are three dimensional organ culture systems containing all the cellular compounds of the original tumor [5]. As the cyto-architecture as well as the extracellular matrix of a tumor are preserved therein we investigated whether PgR remain discernible in this culture system.

Material and Method

25 tumors were analyzed (8 endotheliomatous, 6 atypical, 5 fibroblastic, 4 transitional, 1 psammomatous and 1 microcystic meningioma). The patients (17 females, 8 males) had a mean age of 58.8 + 15.5 years. Native tumor tissue was snap frozen in liquid nitrogen immediately after removal. Additionally, freshly resected tumor tissue was processed further for cell culture: Monolayers were grown in Dulbecco's MEM (+ 10% FCS, 2% glutamine, 100 U/ml penicilline and 100 μg/ml streptomycin) in Lab-Tek 8 chamber tissue culture slides and in 75 cm² plastic flasks. Tumor fragment spheroids were grown on 1% medium agar with liquid medium overlay according to a standardized method previously reported [4,5]. Viable fragment spheroids of 400–500 μg of diameter were chosen for further analysis. Immunohistochemical detection of PgR was performed with rat monoclonal antibodies (Abbott, Wiesbaden, Germany) by the peroxidase-antiperoxidase method with goat-anti-rat secondary antibodies and diaminobenzidine as chromogen. The specimens were counterstained with hematoxilin. Monolayers of the primary passage and passage 3 as well as frozen sections (7 μm) of native tissue and 1 and 3 weeks old spheroids were analyzed.

Results

In 18/25 meningiomas PgR were detectable in native tumor tissue without any correlation to patient's sex or tumor histology (Fig. 1).

A primary passage was obtained in all, a third one in 24/25 specimens. None of the monolayers in either passage depicted PgR (Fig. 2). Cells spreading out of aggregated PgR-positive clumps in primary culture lost PgR-staining when growing as monolayers. Out of 21/25 specimens tumor fragment spheroids could be generated which remained viable for at least 3 weeks. In 8/15 native PgR positive tumors we could detect

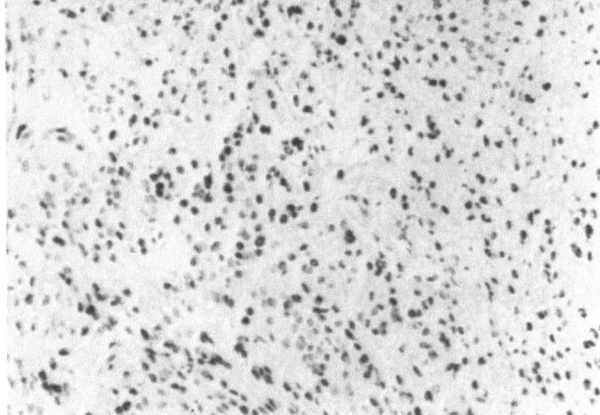

Fig. 1. Immunohistochemical detection of the progesterone receptor (PgR) in native meningioma tissue (dark black perinuclear staining on the microphotography)

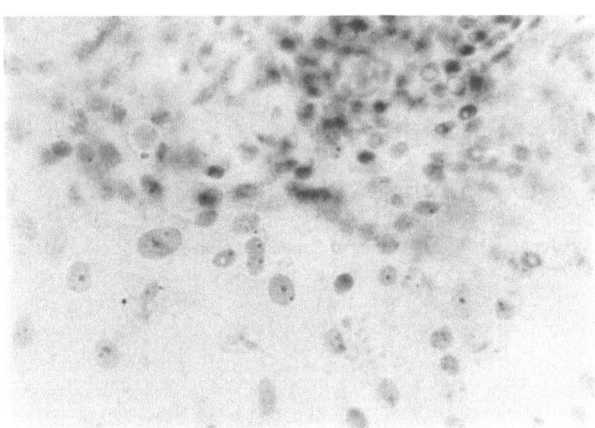

Fig. 2. Absence of PgR staining in cells growing as monolayer (primary passage, same tumor as Fig. 1)

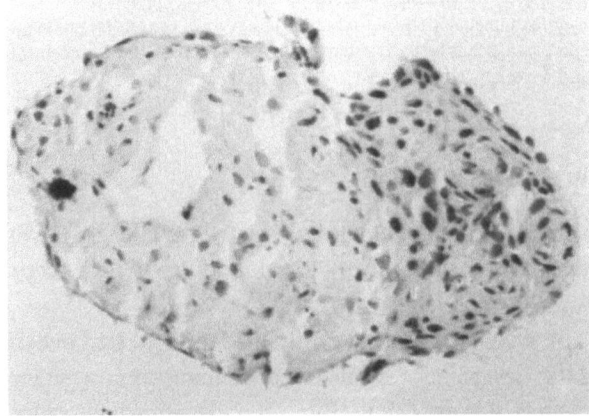

Fig. 3. PgR staining preserved in a 3 weeks old fragment spheroid of the same meningioma as in Figs. 1 and 2. Note the intense black perinuclear staining on the right side of the section

PgR on cryosections of spheroids, this finding remained constant even after 3 weeks of culture (Fig. 3). There was no relation between PgR-staining of spheroids and patient's sex or histological subtypes of the meningioma observable.

Discussion

PgR-expression in native meningioma tissue has been demonstrated by immunohistochemistry and Northern blot analysis [2,6,8,11,15]. Several in vitro studies, however, reported controversial findings in monolayer tissue culture [1,9,13,16]. Thus, the in vitro observation of missing PgR has been interpreted as cell culture artifact, either due to loss of the receptor or overgrowth by a stable, PgR-negative (fibroblastic?) cell population [3,10]. Our data confirm that PgR seems to be lost already in the primary passage of meningioma when cultured as monolayer. As soon as the cells grew out of the cell clumps attached to the bottom of the culture vessel the PgR was no longer detectable. Cells still adherent to each other forming cell clumps depicted strongly positive PgR staining in vitro when the tumor tissue was native PgR positive. One might speculate about the loss of PgR concomitantly to the loss of the three dimensional tissue architecture since overgrowth of a PgR negative cell type seems not to be rather likely in the primary passage. First studies in our laboratory with PgR-negative monolayer cells reaggregated to multicellular spheroids indicate that PgR is indeed re-expressed under the circumstances of three dimensional cellular organisation.

In tumor fragment spheroids, the morphological structure and the cellular complexity of the original tumor tissue remains stable even after longer culture periods [4,5]. Recently we could demonstrate that fragment spheroids of malignant human gliomas are much closer to the in vivo situation in terms of expression of extracellular matrix components and integrin receptors than tumors growing as monolayer [14]. In this context the present data of preserved PgR in tumor fragment spheroids confirm a superior simulation of the in situ status by fragment spheroid culture. As the PgR is preserved in fragment spheroids over at least three weeks of culture we are now going to study whether antiprogesterone-effects might be reliably evaluated in this in vitro model.

Acknowledgements

We highly appreciate the excellent technical assistance of Ms. S. Kerkau and Ms. M. Kapp.

References

1. Adams EF, Schrell UMH, Fahlbusch R, Thierauf P (1990) Hormonal dependency of cerebral meningiomas. Part 2: In vitro effect of steroids, bromocriptine, and epidermal growth factor on growth of meningiomas. J Neurosurg 73: 750–755

2. Blankenstein MA, Blaauw G, Lamberts SWJ, Mulder E (1983) Presence of progesterone receptors and absence of oestrogen receptors in human intracranial meningioma cytosols. Eur J Cancer Clin Oncol 19: 363–370

3. Black P (1993) Meningiomas. Neurosurgery 32: 643–657

4. Bjerkvig R, Tonnesen A, Laerum OD, Backlund EO (1990) Multicellular tumor spheroids from human gliomas maintained in organ culture. J Neurosurg 72: 463–375

5. Bjerkvig R, Hostmark J, Pedersen PH, Laerum OD (1992) Tumor spheroids from biopsy specimens. In: Bjerkvig R (ed) Spheroid culture in cancer research. CRC, Boca Raton, pp 41–56

6. Carrol RS, Glowacka D, Dashner K, Black PM (1993) Progesterone receptor expression in meningiomas. Cancer Res 53: 1312–1316

7. Grunberg SM, Weiss MH, Spitz IM, Ahmadi J, Sadun A, Russel CA, Lucci L, Stevenson LL (1991) Treatment of un-resectable meningiomas with the antiprogesterone agent mifepristone. J Neurosurg 74: 861–866

8. Halper J, Colvard DS, Scheithauer BW, Jiang NS, Press MF, Graham ML, Riehl E, Laws ER, Spelsberg TC (1989) Estrogen and progesterone receptors in meningiomas: comparison of nuclear binding, dextran-coated charcoal, and immunoperoxidase staining assays. Neurosurgery 25: 546–533

9. Jay JJ, MacLaughlin DT, Riley KR, Martuza RL (1985) Modulation of meningioma cell growth by sex steroid hormones in vitro. J Neurosurg 62: 757–762

10. Markwalder TM, Zava DT (1986) Sex steroid hormones and meningioma cell growth. J Neurosurg 64: 341–342

11. Martuza RL, Miller DC, MacLauglin DT (1985) Estrogen and progestin binding by cytosolic and nuclear fractions of human meningiomas. J Neurosurg 62: 750–756

12. Matsuda Y, Kawamoto K, Kiya K, Kurisu K, Sugiyama K, Uozumi T (1994) Antitumor effects of antiprogesterones on human meningioma cells in vitro and in vivo. J Neurosurg 80: 527–534

13. Olsen JJ, Beck DW, Schlechte J, Loh PM (1986) Hormonal manipulation of meningiomas in vitro. J Neurosurg 65: 99–107

14. Paulus W, Huettner C, Tonn JC (1994) Collagens, integrins and the mesenchymal drift in glioblastomas: a comparison of biopsy specimens, spheroid and early monolayer cultures. Int J Cancer 58: 841–846

15. Schrell UMH, Adams EF, Fahlbusch R, Greb R, Jirikowski G, Prior R, Ramalho-Ortigao FJ (1990) Hormonal dependency of cerebral meningiomas. Part 1: female sex steroid receptors and their significance as specific markers for adjuvant medical therapy. J Neurosurg 73: 743–749

16. Waelti ER, Markwalder TM (1989) Endocrine manipulation of meningiomas with medoxiprogesterone acetate: effect of MPA on growth of meningioma cells in monolayer tissue culture. Surg Neurol 31: 96–100

Correspondence: J.C. Tonn, M.D., Department of Neurosurgery, University of Würzburg, Josef-Schneider-Str. 11, D-97080 Würzburg, Federal Republic of Germany.

Acta Neurochir (1996) [Suppl] 65: 108–111
© Springer-Verlag 1996

PET-Study of Intracranial Meningiomas:
Correlation with Histopathology, Cellularity and Proliferation Rate

B. Lippitz[1], U. Cremerius[2], L. Mayfrank[1], H. Bertalanffy[1], R. Raoofi[1], J. Weis[4], A. Böcking[3], U. Büll[2], and J.M. Gilsbach[1]

Department of [1]Neurosurgery, [2]Nuclear Medicine, [3]Pathology, and [4]Neuropathology, Medical Faculty of Technical University Aachen, Federal Republic of Germany

Summary

The glucose metabolism of 62 meningiomas was measured by fluorine -18-2-fluorodeoxyglucose (FDG) PET and correlated with proliferation rate (Ki-67 index) and tumor cellularity. The mean metabolic rate (MRGlu) for meningiomas was 0.26 ± 0.13 mikromol/g/min (range 0.08–0.62 mikromol/g/min).

The relative tumor FDG-uptake (Q-MRGlu) (tumor/contralateral cortex) of all meningiomas was calculated with 0.73 ± 0.37 (0.24–1.79). Differences of Q-MRGlu were significant between the groups with high vs. low cellularity ($p < 0.01$), increased vs. normal proliferation rate ($p < 0.025$) and low (WHO grade I) vs. higher (WHO grades II, III) graded tumors. In recurrent meningiomas (14 tumors) the glucose metabolism was not increased.

The data show that 18 FDG-PET is suitable to serve as non-invasive predictor of tumor growth characteristics in meningiomas.

Keywords: Meningioma; positron emission tomography; grading; recurrence; Ki-67

Introduction

In meningiomas the postoperative recurrence rate is primarily related to location-dependent degree and completeness of resection. Additionally biological factors determine regrowth, as the recurrence rate is increased 5-fold in atypical meningiomas and 10-fold in anaplastic tumors [2]. So far tumor aggressiveness could be quantified only postoperatively using immunohistochemical methods such as proliferation marker Ki-67 or PCNA.

In various other neoplasms it was shown that the glucose metabolism as measured by fluorine-18-2-fluorodeoxyglucose (FDG) PET scan correlated highly with the degree of tumor malignancy.

In the present study the glucose metabolism of 62 meningiomas was measured by PET (FDG) and correlated with the proliferation rate and tumor cellularity in order to define non-invasive and prospective data for the determination of individual growth characteristics and the biological potential for tumor recurrence.

Patients and Methods

Between February 1991 and March 1993, 62 tumors in 60 patients (age range: 23–82 years – mean age 59 years) were studied. The tumor diameter was greater than 2 cm in all cases. Among 62 meningiomas were 14 recurrent tumors (23%).

The PET scan (ECAT 953/15) was carried out 30–60 min after application of 125-330 MBq 18-FDG. In 48/62 studies plasma radioactivity was measured from the arterialized venous blood. For quantification the Sokoloff model [12] and the kinetic constants for FDG determined by Phelps [10] were used. For the lumped constant the value of 0.52, as measured by Reivich [11] for normal brain, was used. Quantification was carried out: 1) as relative FDG-uptake (tumor/contralateral cortex) (Q-FDG), 2) as tumor glucose metabolic rate (MRGlu), 3) as tumor/cortex ratio of the measured metabolic rates (Q-MRGlu).

The Ki-67 proliferation index was quantified in 37 meningiomas using monoclonal antibody MIB-1 (Dianova) in Paraffin embedded tissue: 4000–6000 nuclei were evaluated semi-automatically on a computer screen using the analysis system MIAMED (Leica).

The tumors were reviewed by a neuropathologist. Cellularity was determined in a semi-quantitative manner: "high cellularity" was stated when more than 10 nuclei were found in a 100 mikrometer range.

The data were analyzed according to the Spearman's rank correlation and the U (Mann and Whitney) test.

Results

The mean metabolic rate (MRGlu) for meningiomas was 0.26 ± 0.13 mikromol/g/min (range 0.08–0.62 mikromol/g/min). The mean cortical MRGlu was

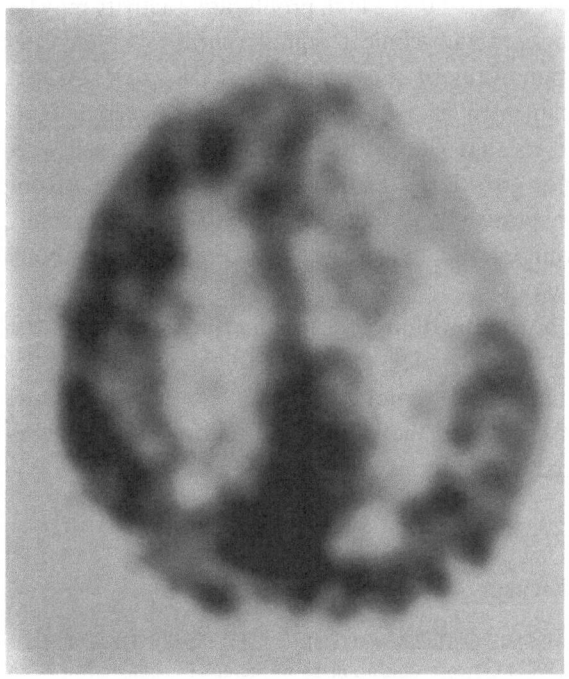

Fig. 1. Frontal meningioma with low glucose metabolism (MR Glu = 0.12 mikromol/g/min; QMRGlu = 0.48

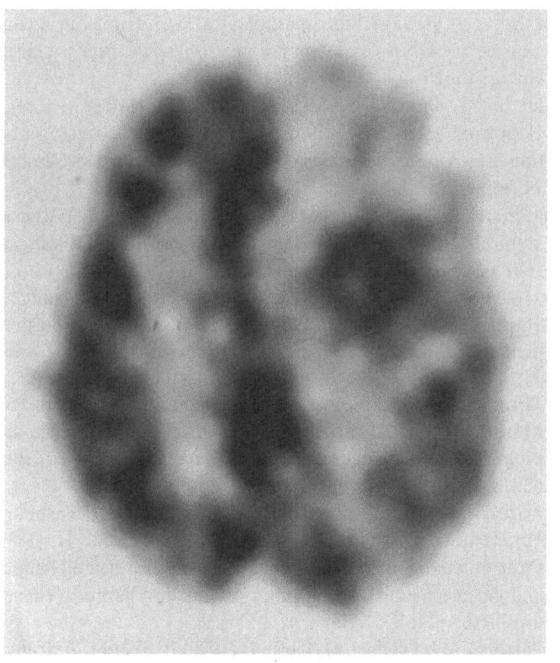

Fig. 2. Atypical frontal meningioma with increased glucose metabolism (MRGlu = 0.5 mikromol/g/min; QMRGlu = 1.47

0.37 ± 0.08 mikromol/g/min. Therefore the relative tumor FDG-uptake Q-FDG (tumor/contralateral cortex) of all meningiomas was calculated with 0.73 ± 0.37 (0.24–1.79) (Figs. 1 and 2).

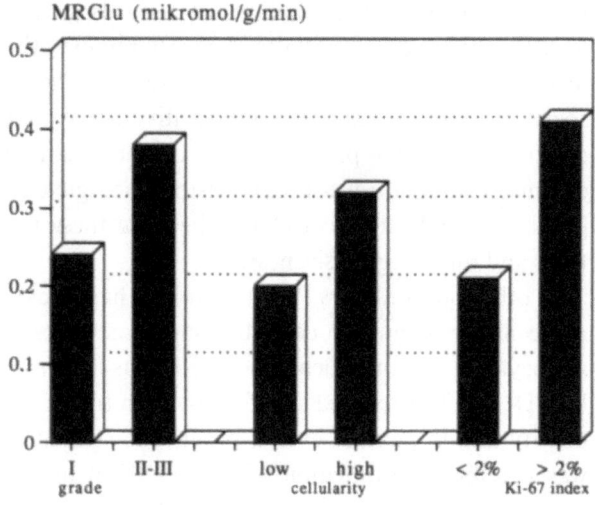

Fig. 3. Significant differences between glucose metabolism (MRGlu): WHO grade I (n=42) vs. grades II, III (n=6), low (n=17) vs. high (n=18) cellularity and Ki-67 index < 2% (n=22) vs. > 2% (n=7)

In recurrent tumors (n=14) glucose metabolism was not increased (MRGlu 0.28 ± 0.14 mikromol/g/min).

The mean Ki-67 proliferation index (n=37) was 1.5 ± 2.6% (range 0–12.7%). In recurrent tumors no increased proliferative activity was detected (1.6% ± 1.8%). 24 (8 out of 14 recurrent tumors) meningiomas showed a higher cellularity.

The influence of cellularity and proliferation rate upon glucose metabolism is shown in Fig. 3. The mean Q-MRGlu (and likewise Q-FDG) of the high cellularity subgroup (n=18) differed significantly from tumors with lower cellularity (n=17 (p<0.01). Likewise significant were differences in Q-MRGlu between meningiomas WHO grade I (n=42) vs. grades II–III (n=6). The proliferation analysis according to the Ki-67 index (n=28) showed a significant correlation (p<0.0025) between Q-MRGlu (and Q-FDG) and proliferation rate.

Discussion

The data demonstrate that FDG-PET is suitable to serve as non-invasive predictor of tumor growth characteristics in meningiomas. Glucose metabolism was increased in high grade meningiomas and in tumors with increased proliferative activity or higher cellularity. In recurrent meningiomas of the present series no elevated metabolism was detected.

So far the increased metabolic or proliferative activity could only be quantified postoperatively by immunohistochemistry or flow cytometry [3]. The preoperative estimation of these factors was not feasible. The present series of meningiomas is the first to show a significant correlation between preoperative metabolic data and histopathological results. The association between cellularity, proliferative activity and glucose utilization is particularly important, since flow cytometric [3] and retrospective histopathologic data [5] likewise indicate a correlation between these features and tumor aggressiveness.

In other brain tumors PET technology has already made major diagnostic contributions. In a series of astrocytomas [4] the glucose metabolism as measured by FDG-PET correlated significantly with tumor cell density and in gliomas even a prognostic estimation was possible by evaluation of PET [9,6]. In meningiomas these parameters had not been analyzed yet.

In the majority of meningiomas the glucose utilization was low when compared with the contralateral cortex. This result is in accordance with other PET studies published before [1,8]. Unlike our data, in DiChiro's group recurrent meningiomas showed an increased glucose metabolism [1]. In 5 out of 11 patients from DiChiro's study multiple meningiomas were found and 5/11 recurrences appeared after 3 or more craniotomies. These features appear to reflect a biologically aggressive subpopulation.

The interpretation of recurrence data in meningioma is difficult, since two distinct mechanisms influence regrowth: tumor location and biological aggressiveness. Retrospective studies documented that tumors in sites with a low percentage of total excisions had a higher regrowth rate [7]. Others demonstrated that an increased cellularity was associated with a higher recurrence rate [5]. Although in meningiomas recurrence/progression primarily depends on factors such as tumor location and completeness of resection and a definite relationship between meningioma subtype and risk for recurrence has not been established [2], certain tumor subtypes seem to show differences in their growth characteristics [5]. Therefore an unselected group of regrowing meningiomas will contain as well "biological" as "location dependent" recurrences. Our finding that the glucose utilization of recurrent tumors was not increased, may reflect the preponderance of "location dependent" recurrences in our group.

The present data demonstrate that the FDG-PET allows to define the subgroup of meningiomas with a higher cellularity or increased proliferative activity, since the measured glucose metabolism showed a significant correlation with the histopathological results and the Ki-67 index. Immunohistochemical

studies suggest that a high proliferative activity may be associated with clinical aggressiveness [3]. After estimation of tumor sizes on repeated CT scans DiChiro documented in a PET study with 17 meningioma patients that glucose utilization correlated well with tumor growth [1]. These results show that data from preoperative PET studies can help to define subgroups of tumors with an increased proliferative or metabolic activity.

The continuation of the present study will show if the PET results match the histopathological criteria [5] with respect of tumor recurrences. For the indications of therapy and the planning of follow-up these additional informations are essential.

References

1. Di Chiro G, Hatazawa J, Katz DA, Rizzoli HV, De Michele DJ (1987) Glucose utilization by intracranial meningiomas as an index of tumor aggressivity and probability of recurrence: a PET study. Radiology 164: 521–526
2. Di Chiro G (1989) Meningioma subtypes: MR and PET features. Radiology 172 (2): 578–579
3. Cruz-Sanchez FF, Miquel R, Rossi ML, Figols J, Palacin A, Cardesa A (1993) Clinico-pathological correlations in meningiomas: a DNA and immunohistochemical study. Histol Histopathol 8: 1–8
4. Herholz K, Pietrzyk U, Voges J, Schröder R, Halber M, Treuer H, Sturm V, Heiss W-D (1993) Correlation of glucose consumption and tumor cell density in astrocytomas. J Neurosurg 79: 853–858
5. Maier H, Öfner D, Hittmair A, Kitz K, Budka H (1992) Classic, atypical and anaplastic meningioma: three histopathological subtypes of clinical relevance. J Neurosurg 77: 616–623
6. Mineura K, Sasajima T, Kowada M, Ogawa T, Hatazawa J, Shishido F, Uemura K (1994) Perfusion and metabolism in predicting the survival of patients with cerebral gliomas. Cancer 73: 2386–2394
7. Mirimanoff RO, Dosoretz DE, Linggood RM, Ojemann RG, Martuza RL (1985) Meningioma: analysis of recurrence and progression following neurosurgical resection. J Neurosurg 62: 18–24
8. Mizukawa N, Hino A, Imahori Y, Tenjin H, Yano I, Yoshino I, Hirakawa K, Yamashita M, Oki F, Nakahashi H (1989) Positron emission tomographic evaluations on hemodynamics and glucose metabolism of brain tumors and perifocal edematous tissues. No To Shinkei 41: 251–258
9. Patronas MJ, DiChiro G, Kufta C (1985) Prediction of survival in glioma patients by means of positron emission tomography. J Neurosurg 62: 816–822
10. Phelps ME, Huang SC, Hoffmann EJ, Selin C, Sokoloff L, Kuhl DE (1979) Tomographic measurements of local cerebral glucose metabolic rate in humans with (F-18)2-fluoro-2-deoxyglucose: validation of method. Ann Neurol 6: 371–388
11. Reivich M, Alavi A, Wolf A, Fowler J, Russell J, Arnett C, MacGregor RR, Shiue CY, Atkins H, Anand A, Dann R, Greenberg JH (1985) Glucose metabolic rate kinetic model parameter determination in humans: the lumped constants and rate constants for (18-F)fluorodeoxyglucose and (11-C) deoxyglucose. J Cereb Blood Flow Metab 5: 179–192

12. Sokoloff L, Reivich M, Kennedy C, DesRosiers MH, Patlak CS, Pettigrew KD, Sakurada O, Shinohara M (1977) The (14-C) deoxyglucose method for measurement of local cerebral glucose utilization: the theory, procedure, and normal values in conscious and anesthetized albino rat. J Neurochem 28: 897–916

Correspondence: Bodo Lippitz, M.D., Department of Neurosurgery, Medical Faculty of the RWTH, Pauwelsstr. 30, D-52057 Aachen, Federal Republic of Germany.

Index of Keywords

SpringerNeurosurgery

Advances and Technical Standards
in Neurosurgery

Volume 22

Edited by L. Symon (Editor-in-Chief), L. Calliauw, F. Cohadon, V. V. Dolenc,
J. Lobo Antunes, H. Nornes, J. D. Pickard, H.-J. Reulen, A. J. Strong, N. de Tribolet

1995. 149 partly coloured figures. XV, 381 pages. ISBN 3-211-82634-3
Cloth DM 298,–, öS 2086,–, approx. US $ 198.00

Advances: K. Thapar, K. Kovacs, E. R. Laws: The Classification and Molecular Biology
of Pituitary Adenomas.
J. P. Chirossel, G. Vanneuville, J. G. Passagia, J. Chazal, Ch. Coillard, J. J. Favre,
J. M. Garcier, J. Tonetti, M. Guillot: Biomechanics and Classification of Traumatic
Lesions of the Spine.
U. Ebeling, H.-J. Reulen: Space-Occupying Lesions of the Sensori-Motor Region.
Technical Standards: J. P. Houtteville: The Surgery of Cavernomas Both Supra-Tentorial
and Infra-Tentorial.
A. Bricolo, S. Turazzi: Surgery for Gliomas and Other Mass Lesions of the Brainstem.
M. Samii, C. Matthies: Hearing Preservation in Acoustic Tumour Surgery.

Volume 21

Edited by L. Symon (Editor-in-Chief), L. Calliauw, F. Cohadon, J. Lobo Antunes,
F. Loew, H. Nornes, E. Pásztor, J. D. Pickard, A. J. Strong, M. G. Yaşargil

1994. 69 partly coloured figures. XIII, 286 pages. ISBN 3-211-82482-0
Cloth DM 228,–, öS 1596,–, approx. US $ 149.00

Advances: G. J. Pilkington, P. L. Lantos: Biological Markers for Tumours of the Brain.
C. Daumas-Duport: Histoprognosis of Gliomas.
F. Cohadon: Brain Protection.
Technical Standards: S. F. Ciricillo, M. L. Rosenblum: AIDS and the Neurosurgeon –
an Update.
M. Choux, G. Lena, L. Genitori, M. Foroutan: The Surgery of Occult Spinal Dysraphism.
P. Cosyns, J. Caemaert, W. Haaijman, C. van Veelen, J. Gybels, J. van Manen, J. Ceha:
Functional Stereotactic Neurosurgery for Psychiatric Disorders: an Experience in
Belgium and The Netherlands.

 SpringerWienNewYork

P.O.Box 89, A-1201 Wien • New York, NY 10010, 175 Fifth Avenue
Heidelberger Platz 3, D-14197 Berlin • Tokyo 113, 3-13, Hongo 3-chome, Bunkyo-ku

SpringerNews

Björn A. Meyerson, Christoph Ostertag (eds.)

Advances in Stereotactic and Functional Neurosurgery 11

Proceedings of the 11th Meeting of the European Society
for Stereotactic and Functional Neurosurgery, Antalya 1994

1995. 72 partly coloured figures. VIII, 139 pages.
Cloth DM 180,–, öS 1260,–, approx. US $ 140.00
Reduced price for subscribers to "Acta Neurochirurgica":
Cloth DM 162,–, öS 1134,–
ISBN 3-211-82720-X
Acta Neurochirurgica, Supplement 64

This is a selection of papers presented at the meeting of the European Society for Stereotactic and Functional Neurosurgery in 1994 and it gives an update of the state-of-art of treatment of movement disorders, pain and stereotactic techniques. The topics include: frameless stereotaxy, the practical usage of a navigator viewing wand system, a novel approach to the localization of the motor cortex, intraoperative monitoring such as microrecording and evoked potentials, the clinical usefulness of magneto-encephalography, a recent study on pallidotomy, a review of the experiences with fetal neurotransplantation for Parkinson's disease, the long-term results and neurophysiological evaluation of baclofen infusion for spasticity, implantation of chromaffin cells in the spinal canal, CT-guided percutaneous cordotomy, spinal cord stimulation applied for "low back pain", a survey of the technique, indications and outcome of nucleus caudalis DREZ operations. Any neurosurgeon interested in stereotactic techniques and treatment of movement disorders and pain will find this book useful as it reflects the most recent advances in the field.

SpringerNeurosurgery

SpringerWienNewYork

P.O.Box 89, A-1201 Wien • New York, NY 10010, 175 Fifth Avenue
Heidelberger Platz 3, D-14197 Berlin • Tokyo 113, 3-13, Hongo 3-chome, Bunkyo-ku

SpringerNews

Wolfgang Koos, Bernd Richling (eds.)

Stereotactic Neuro-Radio-Surgery

Proceedings of the International Symposium
on Stereotactic Neuro-Radio-Surgery, Vienna 1992

1995. 101 partly coloured figures. VII, 119 pages.
Cloth DM 150,–, öS 1050,–, approx. US $ 93.00
Reduced price for subscribers to "Acta Neurochirurgica":
Cloth DM 135,–, öS 945,–
ISBN 3-211-82657-2
Acta Neurochirurgica, Supplement 63

During the last few years stereotactic radiosurgery has become a partner of equal rank within the discipline of neurosurgery. Today it is regarded as being of the same importance as microsurgery and endovascular neurosurgery, branches which have also progressed rapidly in recent years. Breakthrough success, however, requires a combined effort of all partners involved.

The editors have brought together leading experts in the fields of neurosurgery, neuroradiology, neurology, neuropathology, neuroanatomy, radiation oncology, and biophysics to discuss indications and therapeutic strategies in the treatment of arteriovenous malformations and intracranial tumors and to find a common basis for their future work.

SpringerNeurosurgery

SpringerWienNewYork

P.O.Box 89, A-1201 Wien • New York, NY 10010, 175 Fifth Avenue
Heidelberger Platz 3, D-14197 Berlin • Tokyo 113, 3-13, Hongo 3-chome, Bunkyo-ku